BATCH EFFECTS AND NOISE IN MICROARRAY EXPERIMENTS

BATCH EFFECTS AND NOISE IN MICROARRAY EXPERIMENTS

SOURCES AND SOLUTIONS

Edited by

Andreas Scherer
Founder/CEO of Spheromics, Finland

A John Wiley and Sons, Ltd., Publication

This edition first published 2009
© 2009, John Wiley & Sons, Ltd

Registered office
John Wiley & Sons Ltd, The Atrium, Southern Gate, Chichester, West Sussex, PO19 8SQ, United Kingdom

For details of our global editorial offices, for customer services and for information about how to apply for permission to reuse the copyright material in this book please see our website at www.wiley.com.

Wiley also publishes its books in a variety of electronic formats. Some content that appears in print may not be available in electronic books.

Designations used by companies to distinguish their products are often claimed as trademarks. All brand names and product names used in this book are trade names, service marks, trademarks or registered trademarks of their respective owners. The publisher is not associated with any product or vendor mentioned in this book. This publication is designed to provide accurate and authoritative information in regard to the subject matter covered. It is sold on the understanding that the publisher is not engaged in rendering professional services. If professional advice or other expert assistance is required, the services of a competent professional should be sought.

Library of Congress Cataloging-in-Publication Data

Record on File

A catalogue record for this book is available from the British Library.

ISBN: 978-0-470-74138-2 (H/B)

Typeset in 10/12pt Times by Laserwords Private Limited, Chennai, India
Printed and bound in Great Britain by CPI Antony Rowe, Chippenham, Wiltshire.

Contents

List of Contributors

Altman, N, Department of Statistics, Pennsylvania State University, University Park, PA, USA; Naomi@stat.psu.edu

Bao, W, SAS Institute Inc., Cary, NC, USA; Wenjun.Bao@jmp.com

Boedigheimer, M, Amgen Inc., Thousand Oaks, CA, USA; MBoedigh@amgen.com

Bushel, PR, National Institute of Environmental Health Services, Research Triangle Park, NC, USA; Bushel@niehs.nih.gov

Bylesjö, M, Computational Life Science Cluster, Chemical Biology Center, KBC, Umeå University, Umeå, Sweden; Max.Bylesjo@chem.umu.se

Chen, JJ, National Center for Toxicological Research, US Food and Drug Administration, Jefferson, AR, USA; JamesJ.Chen@fda.hhs.gov

Chou, J, National Institute of Environmental Health Sciences, Research Triangle Park, NC, USA; Chou@niehs.nih.gov

Chu, T-M, SAS Institute Inc., Cary, NC, USA; Tzu-Ming.Chu@jmp.com

Coller, JA Jr, Stanford Functional Genomics Facility, Stanford University, Stanford, CA, USA; John.Coller@stanford.edu

Cooper, M, Roche Palo Alto, Palo Alto, CA, USA; Matthew.Cooper.mcl@roche.com

Corton, JC, US Environmental Protection Agency, Research Triangle Park, NC, USA; Corton.Chris@epamail.epa.gov

Deng, MC, Department of Medicine, Columbia University, New York, NY, USA; md785@columbia.edu

Fahlén, J, Department of Statistics, Umeå University, Umeå, Sweden; Jessica.Fahlen@stat.umu.se

Fostel, J, National Institute of Environmental Health Sciences, Research Triangle Park, NC, USA; Fostel@niehs.nih.gov

Freyhult, E, Department of Clinical Microbiology, Umeå University, Umeå, Sweden; Eva.Freyhult@climi.umu.se

Fuscoe, JC, National Center for Toxicological Research, US Food and Drug Administration, Jefferson, AR, USA; James.Fuscoe@fda.hhs.gov

George, NI, National Center for Toxicological Research, US Food and Drug Administration, Jefferson, AR, USA; Nysia.George@fda.hhs.gov

Goodsaid, F, Center for Drug Evaluation and Research, US Food and Drug Administration, Silver Spring, MD, USA; Federico.Goodsaid@fda.hhs.gov

Grass, P, Novartis Institutes of Biomedical Research, Novartis Pharma AG, Basel, Switzerland; Peter.Grass@novartis.com

Hester, S, US Environmental Protection Agency, Research Triangle Park, NC, USA; Hester.Susan@epa.gov

Hong, H, National Center for Toxicological Research, US Food and Drug Administration, Jefferson, AR, USA; Huixiao.Hong@fda.hhs.gov

Hvidsten, T, Umeå Plant Science Centre, Department of Plant Physiology, Umeå University, Umeå, Sweden; Torgeir.Hvidsten@plantphys.umu.se

Johnson, WE, Department of Statistics, Brigham Young University, Provo, UT, USA; Evan@stat.byu.edu

Jones, WD, Expression Analysis Inc., Durham, NC, USA; WJones@expressionanalysis.com

Klebanov, L, Department of Probability and Statistics, Charles University, Prague, Czech Republic; Evbkl@gmail.com

Landfors, M, Department of Mathematics and Mathematical Statistics, Umeå University, Umeå, Sweden; Mattias.Landfors@math.umu.se

Latif, F, Department of Medicine, Columbia University, New York, NY, USA; FL2203@columbia.edu

Li, C, Department of Biostatistics, Dana-Farber Cancer Institute, Boston, MA, USA; CLi@hsph.harvard.edu

Li, JY, Bioinformatics Core, Lineberger Comprehensive Cancer Center, University of North Carolina, Chapel Hill, NC, USA; Jianying_Li@med.unc.edu

Liggett, WS, Statistical Engineering Division, National Institute of Standards and Technology, Gaithersburg, MD, USA; Walter.Liggett@nist.gov

Liu, X, Department of Statistics, Harvard University, Cambridge, MA, USA; XLiu@stat.harvard.edu

Lozach, J, Illumina Inc., San Diego, CA, USA; JLozach@illumina.com

Lucas, AB, Genomics R&D, Life Sciences Solution Unit, Agilent Technologies Inc., Santa Clara, CA, USA; Anne_Lucas@agilent.com

Marron, JS, Department of Statistics and Operations Research, University of North Carolina, Chapel Hill, NC, USA; Marron@email.unc.edu

Mendrick, DL, National Center for Toxicological Research, US Food and Drug Administration, Jefferson, AR, USA; Donna.Mendrick@fda.hhs.gov

O'Lone, R, ILSI Health and Environmental Sciences Institute, Washington, DC, USA; Rolone@hesiglobal.org

Parker, JS, Department of Genetics and Pathology, University of North Carolina, Chapel Hill, NC, USA; ParkerJS@email.unc.edu

Perou, CM, Department of Genetics and Pathology, University of North Carolina, Chapel Hill, NC, USA; CPerou@email.unc.edu

Peterson, RL, Biopharma Consultant, North Chelmsford, MA, USA; Ron.L.Peterson@comcast.net

Pine, PS, Center for Drug Evaluation and Research, US Food and Drug Administration, Silver Spring, MD, USA; P.Scott.pine@fda.hhs.gov

Qian, H-R, Lilly Corporate Center, Eli Lilly and Company, Indianapolis, IN, USA; QianHu@lilly.com

Quackenbush, J, Department of Biostatistics and Computational Biology, Dana-Farber Cancer Institute, Boston, MA, USA; JohnQ@jimmy.harvard.edu

Rustici, G, European Bioinformatics Institute, Wellcome Trust Genome Campus, Cambridge, UK; Gabry@ebi.ac.uk

Rydén, P, Department of Mathematics and Mathematical Statistics, Umeå University, Umeå, Sweden; Patrik.Ryden@math.umu.se

Salit, ML, Multiplexed Biomolecular Science Group, National Institute for Standards and Technology, Gaithersburg, MD, USA; Salit@nist.gov

Scherer, A, Spheromics, Kontiolahti, Finland; Andreas.Scherer@spheromics.com

Schumacher, M, Biomarker Development, Novartis Pharma AG, Basel, Switzerland; Martin.Schumacher@novartis.com

Shahzad, K, Department of Medicine, Columbia University, New York, NY, USA; KhurramShahzadmd@gmail.com

Sharp, FR, MIND, Institute and Department of Neurology, University of California at Davis, Sacramento, CA, USA; Frank.Sharp@ucdmc.ucdavis.edu

Shi, L, National Center for Toxicological Research, US Food and Drug Administration, Jefferson, AR, USA; Leming.Shi@fda.hhs.gov

Sinha, A, Department of Biomedical Informatics, Columbia University, New York, NY, USA; AS2628@columbia.edu

Staedtler, F, Biomarker Development, Novartis Pharma AG, Basel, Switzerland; Frank.Staedtler@novartis.com

Thierry-Mieg, D, AceView Integrative Gene Annotation Group, National Institute of Health, Bethesda, MD, USA; ThierryD@ncbi.nlm.nih.gov

Thierry-Mieg, J, AceView Integrative Gene Annotation Group, National Institute of Health, Bethesda, MD, USA; Mieg@ncbi.nlm.nih.gov

Thomas, R, Hamner Institutes for Health Services, Research Triangle Park, NC, USA; RThomas@thehamner.org

Thompson, KL, Center for Drug Evaluation and Research, US Food and Drug Administration, Silver Spring, MD, USA; Karol.Thompson@fda.hhs.gov

Tong, W, National Center for Toxicological Research, US Food and Drug Administration, Jefferson, AR, USA; Weida.Tong@fda.hhs.gov

Trygg, J, Computational Life Science Cluster, Department of Chemistry, KBC, Umeå University, Umeå, Sweden; Johan.Trygg@chem.umu.se

Walker, WL, MIND Institute and Department of Neurology, University of California at Davis, Sacramento, CA, USA; WLWalker326@gmail.com

Wolfinger, RD, SAS Institute Inc., Cary, NC, USA; Russ.Wolfinger@sas.com

Foreword

It is an honour and a pleasure to introduce this book, as the topic is very dear to my heart. It is no exaggeration to say that, of the countless hours I have devoted in the last decade to puzzling over the analysis of microarray data, I have spent more time on the issues dealt with in this book than on any others. Finding differentially expressed genes, building classifiers, and discovering subclasses are all easy tasks compared with eliminating batch effects. The early gene expression microarray studies were small, say 2–20 arrays, and batch effects were generally not a problem, for the simple reason that the hybridizations could all be done at essentially the same time. This does not mean there were no design issues, for there were, but they seemed and usually were manageable. Naturally even small experiments can be very spread out. I was once asked to help rescue an experiment with only a modest number of arrays, but carried out over 4 years. The attempt failed, and the experiment yielded nothing but very strong year effects. As time went on, and the price of arrays dropped, people started carrying out larger and larger studies. Now it is not uncommon to have studies with hundreds of arrays, which cannot all be handled at once. Samples might come in one at a time, over months or years, but are commonly collected in batches. However the collection, processing and analysis are conducted, time or batch effects are unavoidable. Important though design is, and it is rightly emphasized in the book, there is in general little chance of entirely eliminating these effects. We must do our best with good design, but we must also plan to be in a position to identify and subsequently correct for those effects we are unable to eliminate by design. In this respect, positive and negative controls, and judiciously placed replicates can help a lot. Our job, as always, is to minimize some mixture of bias and variance in our measurements. Every adjustment we carry out will rest on assumptions, and on the suitability of the statistical models we use, but good ancillary measurements can reduce our reliance on these unknowables. The chapters in this first book on batch effects and bias in microarrays are an excellent introduction to the issues, and to the existing solutions. Read on!

Terry Speed
Melbourne
2009

Preface

High-content, high-density long or short oligonucleotide microarrays for simultaneous measurement of redundancy of RNA species are nowadays widely used for hypothesis building in fundamental biology research. They help generate testable ideas on biological processes which have their foundations in gene expression changes. Microarrays are being used more and more in medical research and are tested for their use in personalized medicine, e.g. aiding in disease diagnosis and prognosis. The usefulness of the wealth of data points which are generated by just a single microarray experiment is determined by the quality of the experiment, the data processing, analysis and interpretation. An important aspect in microarray studies is the management of bias. As bias may mask true biological information it can be harmful to the interpretation of the results and thus threaten entire studies. Disregarding sources and magnitude of technical noise and batch effects in the planning phase as well as in the analytical phase may lead to a distortion of the interpretation of results. As one can imagine, this may eventually cause unnecessary waste of time, money and resources and potentially result in misleading conclusions.

This book is addressed to everybody involved in design and/or analysis of microarray studies: researchers, clinicians, laboratory personnel, managers, all those who are responsible for successful quality-driven gene expression studies. Its aim is to increase the awareness of bias in microarray data, describe sources of technical and biological variation in gene expression experiments and genome-wide association studies, suggest ways of reducing bias, and present statistical methods which can estimate bias and alleviate their effects, some of them previously unpublished. The reader will not find comparisons between methods in this book. The purpose of this book is to *present* the issue of technical noise and batch effects in microarray studies, and *show* that it is possible to alleviate such factors to come closer to an interpretation of the relevant biological information.

It was impossible to gather all statistical methods published to date in individual feature chapters. The selection of the methods presented was based on a variety of criteria, almost all of them not scientific. The scope of the book was one, the availability of authors another. Some authors had shifted their research focus since they had published a manuscript which would be relevant to our topic and understandably regretted that they felt they should not contribute. Nevertheless, the editor and authors took great pains that, in addition to the featured methods, other very valuable methods should be described or at least mentioned and the literature cited. I invite the interested reader to continue their reading beyond this book. I am very grateful to all the authors I contacted who considered contributing to this book, whether they eventually did or did not.

Each chapter can be read as a separate, stand-alone entity. The individual writing style and figure design are testament to the experience and knowledge of the teams that contributed. However, through the concept of the book the reader will be guided in steps from recapitulations of technical aspects of gene expression profiling with special attention to technical influences, to experimental design strategies and sophisticated statistical approaches to alleviate batch effects and random and systematic technical noise. The reader is introduced to an emerging idea of mathematical integration of biological noise, before we go on to show that batch effects also occur in genome-wide association studies. Throughout the book the importance of standardization will be stressed, and we conclude by highlighting the value of standard operating procedures in the development of genomics biomarkers, and present an extensive overview of current standardization initiatives. Much of the research presented here has not been published before and can only be found here.

After a high-level introduction to the topic in Chapter 1, we introduce the reader to the practical concept behind some of the most used microarray platforms (Chapter 2), and then stress the importance of experimental design in microarray experiments (Chapters 3 and 4). Chapter 5 points out the impact of batch effects on probe set level and probe level. Chapter 6 estimates the level of variability agreement across five one-color microarray platforms, and analyses the concordance between different normalizations. Chapter 7 presents findings on baseline gene expression variation in laboratory rats, and assesses the contribution of smooth bias and smooth bias correction. Chapters 8 and 9 analyse the extent of batch effects and the impact of normalization in two-dye arrays. Chapters 10 and 11 highlight the empirical Bayes method for batch adjustment, while Chapter 11 stresses the value of identical reference samples in each batch. Chapter 12 introduces principal variance components analysis as a method of batch effect estimation, and Chapter 13 discusses a robust grouped-batch-profile normalization for batch profile estimation, correction, and scoring. Chapter 14 extends the concept of distance weighted discrimination from batch effect correction at the level of the experiment to cross-platform adjustment. Chapter 15 discusses an approach to tackling biological variation, a feature called aggregation. Chapter 16 extends the concept of batch effect from gene expression studies to gcnome-wide association studies, and provides insight into ways of alleviating technical sources of variation in the latter. Chapter 17 focuses on the importance of standard operating procedures in clinical trials. Finally, Chapter 18 is concerned with the efforts of the scientific community toward standardization of procedures, analysis and reporting, in order to make data more comparable.

The editor and the contributors assume a basic knowledge of biological concepts of gene expression and the methods of gene expression analysis.

Certain commercial equipment, instruments, or materials will be mentioned. This does not necessarily imply that those are the best available for the purpose.

Colour images and additional information such as data, scripts can be found on the website for the book, http://www.the-batch-effect-book.org.

All the data sets which are worked on in some of the chapters are publicly available, as are all methods and algorithms used in the book.

I am deeply grateful to all the contributors. All have invested a huge amount of time and effort to help realize this project.

I am indebted to Terry Speed for writing the foreword to this book.

1

Variation, Variability, Batches and Bias in Microarray Experiments: An Introduction

Andreas Scherer

Abstract

Microarray-based measurement of gene expression levels is a widely used technology in biological and medical research. The discussion around the impact of variability on the reproducibility of microarray data has captured the imagination of researchers ever since the invention of microarray technology in the mid 1990s. Variability has many sources of the most diverse kinds, and depending on the experimental performance it can manifest itself as a random factor or as a systematic factor, termed bias. Knowledge of the biological/medical as well as the practical background of a planned microarray experiment helps alleviate the impact of systematic sources of variability, but can hardly address random effects.

The invention of microarray technology in the mid 1990s allowed the simultaneous monitoring of the expression levels of thousands of genes (Brown and Botstein 1999; Lockhart *et al.* 1996; Schena *et al.* 1995). Microarray-based high density/high content gene expression technology is nowadays commonly used in fundamental biological and medical research to generate testable hypotheses on physiological processes and disease. It is designed to measure variation of expression due to biological, physiological, genetic and/or environmental conditions, and it allows us to study differences in gene expression induced by factors of interest, such as pharmacological and toxicological effects of compounds, environmental effects, growth and aging, and disease phenotypes. We note that the term 'variation' describes directly measurable differences among individuals or samples, while the term 'variability' refers to the potential to vary.

Batch Effects and Noise in Microarray Experiments: Sources and Solutions edited by A. Scherer
© 2009 John Wiley & Sons, Ltd

As we shall see in more detail in Chapter 2, relative quantification of gene expression involves many steps including sample handling, messenger RNA (mRNA) extraction, in-vitro reverse transcription, labeling of complementary RNA (cRNA) with fluorescent dyes, hybridization of the labeled cDNA (target) to oligonucleotides with complementary sequences (probes), which are immobilized on solid surfaces, and the measurement of the intensity of the fluorescent signal which is emitted by the labeled target. The measured signal intensity per target is a measure of relative abundance of the particular mRNA species in the original biological sample.

Unfortunately, microarray technology has its caveats, as it is susceptible to variability like any other measurement process. As we will discuss in Chapters 2 and 3, technical variation manifests itself in signal intensity variability. This effect is informally called 'noise': technical components which are not part of the system under investigation but which, if they enter the system, lead to variability in the experimental outcomes. Note that noise is only defined in the context of technology. Since the early years of microarrays, noise and its impact on the reliability of large-scale genomics data analysis have been a much discussed topic. The team of Kerr et al. (2000b) was among the first to recognize the problem and to propose ANOVA methods to estimate noise in microarray data sets. Tu et al. (2002) addressed the issue of how to measure the impact of different sources of noise. Using a set of replicate arrays with varying degrees of preparation differences, they were able to characterize quantitatively that the hybridization noise is very high compared to sample preparation or amplification. They also found that the level of noise is signal intensity dependent, and propose a method for significance testing based on noise characteristics.

The unresolved issue of measurement variability and measuring variability has hampered the great hopes researchers had with the advent of microarray technology and the human genome sequence project. Since consensus technological, analytical, and reporting processes were (and still are) largely missing, it appeared that not only were gene expression data irreproducible, but also the results were very much dependent on the choice of analytical methods. A lively discussion on the validity of microarray technology resulted in publications and comments like 'Microarrays and molecular research: noise discovery?' (Ioannidis 2005), 'An array of problems' (Frantz 2005), countered by 'Arrays of hope' (Strauss 2006), and 'In praise of arrays' (Ying and Sarwal 2008), and publications which raise questions about the reproducibility of microarray data (Marshall 2004; Ein-Dor et al. 2006) or showing increased reproducibility (Dobbin et al. 2005b; Irizarry et al. 2005; Larkin et al. 2005).

Shi et al. addressed this issue in a systematic manner and in 2006 published a comparative analysis of a large data set which had been generated by MicroArray Quality Control Consortium (MAQC) with 137 participants from 51 organizations (Shi et al. 2006). The data set consists of two commercially available RNA samples of high quality – Universal Human Reference RNA (UHRR) and Human Brain Reference RNA – which were mixed in four titration pools, and whose mRNA levels were measured on seven microarray platforms in addition to three alternative platforms. Each array platform was deployed at three test sites, and from each sample five replicates were assayed at each site. This information-rich data set is an excellent source for the investigation of technological noise, and some of its data will be used in a number of chapters in this book. The project showed that quantitative measures across all one-color array platforms had a median coefficient of variation (CV) of 5–15%, and a concordance rate for the qualitative calls ('present', 'absent') of 80–95% between sample replicates. Lists of differentially expressed genes overlapped by about 89% between the test sites using the same platform, dropping to 74% overlap across platforms. The important conclusion the authors made is that the performance of the microarray technology in their study speaks for its use in basic research and may lead to its use as clinical diagnostic tool as well. The authors further point out that standardization of data reporting, analysis tools and controls is important in this process.

As pointed out earlier, 'noise' is used to informally describe measurement variability due to technical factors. In the context of biological variability, the term 'noise' will be avoided in the course of this book. Here we suggest the use of the more neutral term 'expression heterogeneity'. The basis of expression heterogeneity lies within the inherent differences in the nature of the subjects or specimen which are studied. It is dependent on the subjects' physiological states, their gender, age, and genetic aspects (Brem *et al.* 2002; DeRisi *et al.* 1997; Rodwell *et al.* 2004). Variability due to biological factors cannot be avoided or minimized and may sometimes even be useful and important. To minimize technical and biological variability, animal toxico- and pharmacogenomic studies are performed under standardized conditions until the tissue harvest (and further): housing in standardized cages, gender- and age-matching, and technical processes which adhere to standard operating procedures and good laboratory practice. However, one or more animals may react differently to the treatment than the others, and their expression signature may indeed provide very valuable information for the investigators. In another example, measuring gene expression heterogeneity is important in gaining understanding pathogenesis in the concept of personalized medicine (Anguiano *et al.* 2008; Bottinger *et al.* 2002; Heidecker and Hare 2007; Lee and Macgregor 2004).

In contrast to the random nature of 'noise', the nature of 'batch effect' is systematic. The term 'batch effect' refers exclusively to systematic technical differences when samples are processed and measured in different batches. Lamb *et al.* were confronted with batch effects when they tested 164 small molecules in cell culture. Since not all cells could be grown at the same time due to the large amount of cells they needed, the cells had to be grown in batches. Hierarchical clustering showed that this batch effect masked the more subtle effects of small-molecule treatment (Lamb *et al.* 2006). As we shall see in other examples and sources of batch effects in the course of this book, batch effects can virtually be generated at each step of an experiment, from sample manipulation to data acquisition. They are unrelated to the biological, primary modeled factors. Batch effects introduce system variability which can be of confounding nature and mask the outcome. Proper evaluation of sources and potential magnitude of technical noise during the planning, execution and analytical phase helps in extracting relevant biological information.

A wider term describing not only technical but also other aspects of confounding the data is 'bias'. We speak of bias where one or more systematic factors affect one or more experimental groups but not all. Bias may be defined as unintentional, systematic erroneous association of some characteristic with a group in a way that distorts a comparison with another group (Ransohoff 2005). There are different types of bias, among them the following: selection bias, when, for instance, the control population has less potential for exposure than the cases; self-selection bias, when only a certain, not-representative subpopulation serves as voluntary study population; measurement bias, due to systematic differences in the measurement process; and cognitive bias, where the decision is based on educational history.

Manageable potential sources of batch effects and bias should be accounted for during the experimental design phase. They should be as consistent as possible throughout the experiment. Monitoring these sources and reporting of deviations from the standard is detrimental. In Chapters 3 (by P. Grass) and 4 (by N. Altman) it will be shown that thoughtful experimental design can alleviate the impact of batch effects. Randomization and blocking are two concepts which accomplish this. Randomization is a concept in which experimental units (e.g. samples) are assigned to groups on a strictly random basis. This means that every sample has the same chance of being selected, and that the sample is representative of its study group. Blocking is a strategy of grouping samples into experimental units which are then homogeneous for the factors which are studied. This is important when samples cannot be processed on a single day. As in the case of Lamb *et al.*, growing all cells destined to be control cells on one day and growing all treated cells on another would introduce a confounding time effect. Lamb *et al.* (2006) carefully chose a setting where treated cells were grown on the same plate as the corresponding control cells.

Chapters 5 through 15 deal with descriptive and analytical ways of exploring the nature, extent, and influence of batch effects, in addition to providing statistical means of adjusting confounding effects. As the MAQC project has stressed, standardization of data acquisition, analysis and reporting is an important factor in making gene expression studies transparent and comparable. This is further highlighted by Frueh (2006), who points out the necessity of a 'best microarray practices' strategy to ensure quality of starting material, data, analysis, and reporting, and interpretation. In this book Shahzad *et al.* (Chapter 17) will show the benefit of the application of standard operating procedures in the development of a commercial genomics biomarker panel. The book will close with an overview of the status of various initiatives which are currently developing standardized procedures for biomedical research (Chapter 18, Rustici *et al.*).

The purpose of the book is to raise the awareness of sources of variability in microarray data, especially of batch effects and bias. It should serve as guidance and starting point for further studies at the same time. Biologists and managers who plan microarray studies are invited to read the book, as well as laboratory personnel, statisticians, and clinicians who execute the study and analyse the data.

2

Microarray Platforms and Aspects of Experimental Variation

John A Coller Jr

Abstract

In the early 1990s, a number of technologies were developed that made it possible to investigate gene expression in a high-throughput fashion (Adams *et al.* 1991; Liang and Pardee 1992; Pease *et al.* 1994; Schena *et al.* 1995; Velculescu *et al.* 1995). Of these technologies, the DNA microarray has become the standard tool for investigating genome-wide gene expression levels and has revolutionized the way scientists investigate the genome. This chapter will give a brief overview of current microarray technologies that are being used today. The chapter will also go through a typical experimental procedure discussing various sources of experimental variability and bias that may affect the data generated and experimental results.

2.1 Introduction

In simple terms, a DNA microarray is a patterned array of immobilized nucleic acid constructs, called probes, that have been constructed in situ or mechanically deposited onto a flat solid substrate, usually glass or silicon. The microarray is used to interrogate the amount of messenger RNA (mRNA), or expressed genes, present in the biological sample under investigation through a hybridization experiment. Prior to the completion of the Human Genome Project (Lander *et al.* 2001; Venter *et al.* 2001), whole-genome expression arrays used complementary DNA (cDNA) probes as the immobilized nucleic acid constructs. However, the current trend is to use the information from genome sequencing projects to design oligonucleotide probes to manufacture microarrays.

Currently, there are four dominant microarray vendors offering oligonucleotide microarrays, with a large number of other vendors offering either competing or supportive products. These vendors are Affymetrix (Santa Clara, CA, USA), Agilent Technologies

Batch Effects and Noise in Microarray Experiments: Sources and Solutions edited by A. Scherer
© 2009 John Wiley & Sons, Ltd

(Santa Clara, CA, USA), Illumina (San Diego, CA, USA), and Roche Nimblegen (Madison, WI, USA). Even though each vendor uses a slightly different method to manufacturer microarrays, the underlying mechanism of use is the same. That is, a nucleic acid hybridization reaction is used to capture and distinguish one nucleic acid sequence from another. We will briefly discuss the various microarray manufacturing technologies in the next section, before discussing experimental issues in Section 2.3.

Due to the nature and general applicability of the hybridization reaction between two nucleic acid molecules, many scientific applications have been developed using the microarray. Some of these applications include: gene expression profiling (Brown and Botstein 1999), comparative genomic hybridization (CGH) (Pollack *et al.* 1999), single nucleotide polymorphism (SNP) detection (Hacia *et al.* 1999), and protein-nucleic acid interaction detection (Liu *et al.* 2005). In all of these applications, the basic fundamental experimental process is the same: obtain DNA or RNA from a biological source of inter-est, hybridize it to the probes on the microarray, and measure the amount of hybridized material. Details of how the nucleic acid material is obtained and how the amount of hybridized material is measured will be discussed later for gene expression. Independent of the particular application, a microarray is a general tool for capturing nucleic acids with complementary nucleic acid probes that have been immobilized onto a surface.

2.2 Microarray Platforms

Current microarray manufacturing methods can be separated into four categories: in-situ synthesis of nucleic acid probes using photolithographic techniques; in-situ synthesis using ink-jet or noncontact printing technology; robotic deposition of prefabricated nucleic acids using contact or noncontact printing technologies; and randomized placement of prefab-ricated nucleic acids attached to beads. This section discusses each of these fabrication methods by describing the microarray platforms using the technology.

2.2.1 Affymetrix

The Affymetrix platform makes use of technology that was developed in the semiconduc-tor industry for fabricating microarrays known as GeneChips® (Fodor *et al.* 1991). The technology, called photolithography, uses a set of opaque masks to selectively block or expose light to regions of a solid support substrate, called a wafer. By selectively exposing portions of the wafer to ultraviolet light, it is possible to synthesize nucleic acids or other molecules through photochemical directed reactions in a spatially dependent manner.

Affymetrix typically synthesizes 25-mer oligonucleotide probes on their arrays. The oligonucleotides are synthesized by repeating a couple of chemical reaction steps. In the first step, a wafer is exposed to near-ultraviolet light that removes photolabile protecting groups on the ends of the oligonucleotide probes. Only oligonucleotides on the portion of the wafer that is exposed to ultraviolet light, not blocked by the opaque mask, is deprotected. These exposed portions are now activated for the next step in the process. The oligonucleotides on unexposed regions of the wafer remain protected and are there-fore inactive. In the second step, or coupling reaction, deoxynucleoside phosporamidite monomers with photolabile protecting groups are washed over the solid support surface.

The monomers react only with hydroxyl groups that have been exposed in the previous step. Only one type of monomer, A, G, T, or C, is added per cycle. The photolithographic mask is then exchanged for deprotecting the next group of nucleotides that need to be added to the growing strands of oligonucleoide probes. Excess reagents are removed after each step and the procedure is repeated until all oligonucleotide probes for an array have been synthesized. Many arrays can be manufactured on the same wafer.

The fidelity of synthesis for in-situ synthesis is important to consider. Incomplete deprotection or coupling can lead to sequence differences between probes on arrays within or between batches. Each gene on an Affymetrix array typically consists of multiple probe pairs. Each pair consists of two oligonucleotide sequences: one that is complementary to the transcript under investigation, called a perfect match (PM) probe, and one that has a central mismatch (MM) in the probe sequence. The mismatch probe is used to correct for and discriminate nonspecific hybridization events and calculate background. An expression value is calculated using all the probes for a given gene.

2.2.2　Agilent

Agilent Technologies fabricates microarrays using ink-jet technology that was originally developed for the printing of dyes or inks on paper (Hughes *et al.* 2001). Unlike Affymetrix's use of photolithography to spatially direct oligonucleotide synthesis, Agilent uses modified ink-jets to spatially deliver and separate chemical reagents on a glass substrate. This technology is very flexible since a computer file instead of a set of physical masks is used to define the pattern and sequences of oligonucleotides on the microarray.

Even through the essential reaction steps of coupling followed by decoupling are similar, the in-situ chemical synthesis of oligonucleotides using Agilent technology is slightly different than that described above for Affymetrix. First, nucleic acid monomers are deposited onto spacially distinct sites on a glass substrate using ink-jets. Each site receives one of four possible modified nucleic acid precursors, A, T, C, or G. The deposited precursors react with exposed reactive sites on the growing oligonucleotide. After the reaction, the excess monomers are washed away. The substrate is then bathed in an acid to deprotect the last nucleotide precursor that was added during the last step. The oligonucleotide is now activated and ready for the addition of the next A, T, C, or G precursor deposited by the ink-jets.

It is very easy to modify the oligonucleotide sequence of any or all of the probes on an array with this technology. With the flexibility of the ink-jet technology it is possible to empirically design probes against genes of interest.

2.2.3　Illumina

Unlike the in-situ fabrication techniques of Affymetrix, Agilent, and Nimblegen, Illumina uses a self-assembly method to construct microarrays, called BeadChips, using the random placement of bulk synthesized oligonucleotide probes onto a patterned substrate. In this method, oligonucleotide probes of length 70 are first synthesized using Illumina's Oligator technology, which uses a centrifugation technique to produce large quantities of oligonucleotides (Lebl *et al.* 2001). Each oligonucleotide probe contains a sequence of 50 nucleotides for the gene being interrogated and a sequence of 20 nucleotides for determining the bead position on the array. The oligonucleotides, which are covalently attached to silica beads, are pooled and placed onto a substrate with etched wells. The beads within each pool randomly find individual wells on the substrate to sit. Due to the randomness of the distribution of probes on the microarray, each array created has a different distribution pattern and number of oligonucleotide probes. On average, each array contains approximately 30 beads of each probe type. Illumina also makes other array formats (see http://www.illumina.com for details).

Unlike the other microarray platforms discussed, each Illumina array is unique and requires the position of beads to be extracted before it can be used. To determine the distribution of beads on the array, each microarray goes through a series of hybridization reactions to deconvolute the pattern of probes on the substrate (Gunderson *et al.* 2004). During each step in the process, two fluorescent targets are hybridized to the array, imaged, and then removed from the array for the next hybridization reaction with a different set of fluorescent targets. After each reaction, each probe has one of three states based on the sequence of the two targets being hybridized, red, green, or no color. During the next hybridization reaction, each probe will have another set of three possible states based on a different set of target sequences. With an appropriate set of fluorescent targets designed against the 20-mer probe sequences, each bead or probe can be uniquely identified on the array.

Since the placement of probes on each Illumina microarray is unique, biases due to where the probes sit on the array, i.e. center or edge, are eliminated. Each probe on each array is also experimentally tested during the decoding hybridization reactions, which gives added quality control information for the manufacturing of the arrays and probes.

2.2.4   Nimblegen

Nimblegen uses a photolithographic technique to fabricate microarrays similar to that of Affymetrix (Singh-Gasson *et al.* 1999). However, Nimblegen does not use a set of physical masks to direct the photochemical reactions. Instead, Nimblegen uses an array of micromirrors to selectively direct light onto regions of the microarray substrate. Each micromirror on the array either reflects light onto a specific region of the substrate or does not reflect light, creating a digital mask. Since the digital micromirror array is controlled by software, it is easy to change oligonucleotide sequences. The steps in fabricating the arrays are similar to those for Affymetrix

2.2.5   Spotted Microarrays

Spotted microarrays (Schena *et al.* 1995) are typically fabricated using a robot to place pre-synthesized oligonucleotides or cDNA onto a solid support, usually a functionalized glass microscope slide. The robot is outfitted with a printing head with pins that pick up material from a microtiter plate and transfers the material to a slide. The pins within the print head use capillary action to pick up and deposit the material similarly to the way a quill pen picks up and deposits ink on paper. Microarrays are fabricated by repetitively picking up material from the microtiter plates and placing this material onto slides in a rectilinear pattern.

2.3 Experimental Considerations

The purpose of any microarray expression experiment is to measure the abundances of RNAs in a biological sample in a particular physiological state. Therefore, it is important that the steps taken during the experimental process do not cause measured RNA levels to deviate from levels in the actual state of interest. Any processes or procedures that may cause an experimental bias in the measured expression values should be reduced or eliminated if possible. The adage from computer science, 'garbage in, garbage out', certainly applies in the case of microarray experiments. If the material put onto a microarray does not accurately represent the starting RNA material, the data obtained will be of little value or worthless even though the data may appear to be perfectly usable.

Below we will go through the steps involved in a typical microarray experimental procedure, from the acquisition of a biological sample through to data extraction. The steps involved include extracting RNA from a biological sample, amplifying the extracted RNA, labeling the amplified product for detection, hybridization to the microarray, and finally scanning and extracting the data. Many of the experimental issues that can lead to systematic biases in the data are summarized in Table 2.1. The issues that are discussed below in no way represent a complete list of all sources of systematic variation that may impact a microarray experiment. However, those listed below and discussed elsewhere in this book should serve as a place to start.

2.3.1 Experimental Design

Prior to beginning any microarray experiment, it is important to thoroughly consider the design of the experiment so that the questions being investigated can be adequately answered. Potential confounding factors need to be dealt with in the initial design stages so that any unnecessary experimental bias does not obscure the experimental results (Lee *et al.* 2005; Rothman *et al.* 1980; Yang and Speed 2002). Inherent biological differences between samples, such as gender or age, can mask the differences between samples or treatments being sought. Issues involved in the experimental design of a microarray study will be discussed in depth in the following chapter.

2.3.2 Sample and RNA Extraction

Prior to extracting RNA from a biological sample, samples need to be acquired and stored in an appropriate manner to preserve RNA integrity and abundances. One way to accomplish this is to snap-freeze samples in liquid nitrogen at the time of collection. If tissue or cells are not immediately frozen, they may begin to differentiate from their physiological state of interest. Significant gene expression differences can occur from ischemia when surgically extracted tissue is allowed to sit for varying times after extraction or during procedures (Huang *et al.* 2001; Lin *et al.* 2006). Additionally, RNA in samples that are not

Table 2.1 Sources of experimental variation during microarray studies.

Source of variation	Effect	Solution	Literature
Improper experimental design			
Epidemiological (e.g. age, gender, etc.)	Masking of physiological state under investigation	Balanced experimental design	Lee et al. (2005) Rothman et al. (1980)
Sample and RNA extraction			
Tissue heterogeneity	Masking of tissue or cell population under investigation	Cell sorting Tissue dissection Laser micro-dissection	Bakay et al. (2002)
Temporal and biological variation in expression	Masking of the biological state under investigation	Balanced experimental design	Whitney et al. (2003) Boedigheimer et al. (2008)
Expression changes after tissue extraction (ischemia, shock, etc.)	Measured RNA abundances different than true physiological state being studied	Proper extraction and methods to eliminate change, such as freezing, stabilization, and lysis	Huang et al. (2001) Lin et al. (2006)
Degraded RNA	Measured RNA abundances different than true physiological state being studied		Thompson et al. (2007)
RNA processing			
Amplification biases	RNA abundances change with different protocols and handling	Strict adherence to protocols and standard operating procedures	Ma et al. (2006) Boelens et al. (2007)
Labeling biases	Measured signals differ from actual abundances and are dependent on actual procedure used	Strict adherence to protocols and standard operating procedures	Lynch et al. (2006)

Hybridization Nonuniform hybridization	Spatial signal biases and nonuniform high backgrounds	Use of mixing hybridization system and optimized hybridization conditions	Schaupp et al. (2005)
Washing Cy5 degradation	Cy5 molecule degrades under ozone exposure	Minimize or eliminate ozone during washing and other experimental steps	Fare et al. (2003) Branham et al. (2007)
Scanning System stability	Variation in signal outputs	Maintain scanner calibration and operating parameters	Leo (1994) Satterfield et al. (2008)
System settings	Scan-to-scan variability	Maintain system settings and calibration for all scans	Shi et al. (2005b)
Clinical diagnosis Subjective analysis of specimen	Systemic bias in assessment due to single or multiple pathologists making diagnosis	Single pathologist or institution assessing specimens	Furness et al. (2003) Daskalakis et al. (2008)
Data interpretation Selection bias	Bias in selecting data sets for training and validation	Randomization of selections, cross-validation, and bootstrapping	Ambroise and McLachlan (2002)

immediately frozen may also begin to get chopped up by RNases. This degradation can cause experimental biases in the resulting experimental data (Thompson *et al.* 2007). If samples cannot be frozen immediately, they should be lysed and stabilized in an RNase inhibitor, such as RNAlater (Ambion Inc., Austin, TX, USA), as soon as possible to prevent degradation of RNA by RNases.

Samples that have been surgically extracted usually consist of heterogeneous popula-tions of cells and tissues. Cells from the surgical margin of extracted tissue may dominate cell populations and mask the gene expression signatures and information sought (Bakay *et al.* 2002). Methods such as cell sorting, tissue dissection, and laser micro-dissection should therefore be considered to eliminate such confounding sample collection biases if possible. Cell populations can also change over time or due to physiological state (Whitney *et al.* 2003; Boedigheimer *et al.* 2008), so the time when samples are collected should be considered during the design phases of the experiment.

The extraction of RNA from a biological sample is a critically important step since the quality of the starting material used in any microarray experiment determines the quality of the resulting data. The guanidine-based TRizol extraction method (Chomczynski and Sacchi 1987) is the method most often used to extract RNA for microarray experiments. Care must be taken so that the RNA extracted is free from contaminants such as proteins, polysaccharides, and DNA. If contaminants are not removed they may affect downstream processes such as enzymatic reactions and array hybridization.

RNA quality and integrity should always be evaluated prior to continuing with any microarray experiment. First, the quantity and purity of the RNA should be measured with a spectrophotometer. Good quality RNA will typically have a 260 nm to 280 nm ratio of 1.8 or larger. Lower ratios may indicate that the RNA is contaminated with protein or other contaminant absorbing at 280 nm. Quality assessment is usually done using an ND-1000 spectrophotometer (Thermo Fisher Scientific, Waltham, MA, USA) or other spectrophotometer to measure the quantity and purity of RNA. Integrity measurements are usually done with an Agilent 2100 Bioanalyzer (Agilent Technologies, Santa Clara, CA, USA) or agarose gel. The Bioanalyzer calculates an RNA integrity number (RIN) for eukaryotic total RNA based on the 18S and 28S ribosomal RNA (rRNA) peaks. The RIN number is used to give an indication of the integrity of the RNA (Schroeder *et al.* 2006). However, the electropherogram should always be examined to verify that the RNA is good. If the extracted RNA is not of high quality, purity, and integrity, it may be questionable whether to continue to the next step in the experimental process.

2.3.3 Amplification

Ideally, each experiment should start with sufficient material so that the mRNA can be directly labeled and measured on the microarray. However, for many samples and experiments, this criterion is usually not satisfied. Therefore, some method of amplifying the starting RNA material is necessary that faithfully reproduces the relative abundances of mRNAs present in the samples of interest. The Eberwine process is commonly used to amplify mRNA targets for DNA microarrays (Van Gelder *et al.* 1990).

A typical amplification process begins with first strand synthesis of cDNA from mRNA by priming with a modified oligo(dT) or random primer and extending with reverse transcriptase. The primer incorporates a T7 or other bacteriophage promotor on the 5′ end. After first round synthesis, second strand synthesis generates double-stranded DNA by extending the first round cDNA with DNA polymerase. T7 polymerase is then used to synthesize antisense RNA (aRNA). This process can be repeated using aRNA as a template for a second round of amplification. Each step in the amplification process can introduce biases due to differences in RNA length, sequence, or secondary structure.

Any variations in the methods or protocols used to amplify sample RNA can lead to systematic differences between amplified products. For example, if either the incubation times, temperatures, or enzymes used in the reactions are modified, the relative ratio of mRNA species may not reflect the initial sample abundances (Ma *et al.* 2006). Also, since the enzymes used in amplification reactions usually start at one end of the RNA templates, there can be a signal dependence on the 3′ distance of the probes on the microarray (Boelens *et al.* 2007). Care should be taken to make sure that protocols and procedures are adhered to at all times and the processing of samples should be distributed such that day-to-day or technician-to-technician biases do not confound results.

2.3.4   Labeling

To get a measurement of the RNA abundances in the original sample, a way of quantifying the amplified material generated in the previous step is necessary. To generate a measurable signal, a fluorescent molecule is usually incorporated into the aRNA. This is usually done by reverse-transcribing the aRNA into cDNA with random primers in the presence of fluorescent nucleotides. Methods incorporating biotinylated nucleotides during the amplification process followed by staining with avidin conjugated fluorescently labeled antibody (Ab) have also been used to label samples.

The method used to label samples can have an effect on measured signals (Lynch *et al.* 2006). If the labeling efficiency is high and too many dye molecules are incorporated such that they are close enough to quench each other, the measured signal may actually decrease instead of increase. The quenching of the cyanine dyes, Cy3 and Cy5, is nonlinear and the phenomenon will systematically bias calculated expression levels ('t Hoen *et al.* 2003).

2.3.5   Hybridization

The hybridization of labeled samples on a microarray is similar to that of conventional Northern and Southern assays except that the probe is immobilized and the target is in solution. The basic process is quite simple and involves incubating at high temperature for a given length of time. Initial hybridization methods used cover glass and relied purely on diffusion of targets to locate probes on the array, limiting signals due to the slow rate of diffusion of targets. Additionally, these hybridization methods were sensitive to evaporative effects, resulting in nonuniform and high backgrounds. Currently, most hybridization methods incorporate active mixing of the hybridization solution. Mixing of the hybridization solution increases signal intensities and improves slide-to-slide replicate correlations (Schaupp *et al.* 2005). Agilent introduces a small air bubble into the chamber used for hybridization. The air bubble mixes the hybridization solution during rotation of

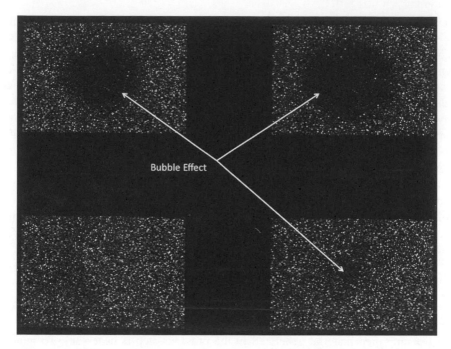

Figure 2.1 Cropped image from Agilent 8-plex array displaying four identical arrays. Three of the arrays did not receive enough hybridization buffer and resulted in varying amounts of data loss.

the chamber in a hybridization oven. Due to the rotational symmetry of the system, the bubble generates a systematic bias in signal intensities around the center of the array and needs to be accounted for by the extraction software. Identical probes distributed across the microarray are used to adjust for this effect. In cases where the volume of the bubble is too large, the bias cannot be corrected. Figure 2.1 illustrates this problem. Hybridization ovens can lose calibration abruptly or over time. Therefore, it is necessary to make sure that the oven is performing properly at all times. Since the length of time that a microarray spends in the hybridization oven is usually greater than 12 hours, placing the oven on emergency power or uninterruptable power supply (UPS) is recommended.

2.3.6 Washing

After hybridization, it is necessary to wash the microarray prior to scanning. The purpose of washing the microarray is to remove any labeled targets that have not hybridized specifically to their conjugate probes on the microarray. The washing process usually involves a number of different washes of different stringencies and a final step to dry the microarray. Inherent inconsistencies in the washing and dry steps can cause large variability in measured signals and background. Additionally, environmental ozone (O_3) can degrade cyanine dyes during these steps (Fare *et al.* 2003; Branham *et al.* 2007). Cy5 is more susceptible to degradation than Cy3 and results in systematic biases when calculating log ratios for two-color experiments and usually cannot be removed by standard normalization schemes.

A number of experimental precautions can be taken to reduce the systematic effects of the washing and drying steps. First, standard operating procedures (SOPs) should be developed and strictly followed so that any variations in how the microarrays are washed and dried are reduced. Second, if possible, procedures should be done in a reduced ozone environment to eliminate degradation of the Cy5 molecule. Methods for reducing laboratory ozone levels are discussed in the above references. Alternatively, dyes that are not as sensitive to ozone degradation can be used.

2.3.7 Scanning

To measure the amount of fluorescent dye molecules that have been incorporated into the material hybridized to the array, the fluorescence is typically measured using a microarray scanner. Scanners usually have two forms: a laser based scanner or a charged-coupled device based imaging system. Both systems have advantages and disadvantages; we will concentrate on laser based scanners since these are more common.

The microarray scanner works by scanning a laser beam across the microarray and recording the fluorescent emission with a photomultiplier tube (PMT). The spatial resolution is typically better than 5 μm for present-day instruments. Excitation and emission filters typically filter out light not at the right wavelength for the fluorescent dyes used. The PMT converts photons that hit a photocathode into electrons that are subsequently amplified through a set of electrodes, called dynodes. The amplified electrons are collected by an anode and then further amplified and digitized by an electronic system. The measured fluorescent signal is assumed to be directly proportional to the amount of hybridized material to the probes of the array.

A number of factors cause variation in the data produced when scanning a microarray. If the laser system used to excite the fluorescent dye molecules is unstable, the resulting signals will vary with the fluctuating excitation. It is usually recommended to wait for the lasers to stabilize before starting a scan. Detection systems can also be unstable over short and long time periods and result in data variation (Leo 1994). The scanner should therefore be calibrated on a regular basis to minimize stability drift (Satterfield *et al.* 2008). PMT gains can also be increased or decreased by changing the PMT voltages. It must be recognized that when the gain of the PMT gain is changed, its calibration curve changes (Shi *et al.* 2005b). This implies that measured RNA abundances will shift with changing voltages. It is therefore important to maintain scanner calibration and adhere to identical settings when scanning. PMTs are extremely sensitive photo-detection systems and can amplify single photons, so light leaks in the scanner housing can cause additional sources of noise. Figure 2.2 illustrates this effect.

The cyanine dyes that are usually used as fluorescent reporters, Cy5 and Cy3, are not completely stable, degrade upon ozone exposure, and overlap slightly in their excitation and emission wavelengths. Care must be taken to minimize exposure of hybridized arrays to minimize exposure to light, high temperatures, and high ozone levels even after washing. If there is a chance that the arrays may need to be scanned at a later time, they should be stored in a cool place in a light-tight enclosure under vacuum or inert gas.

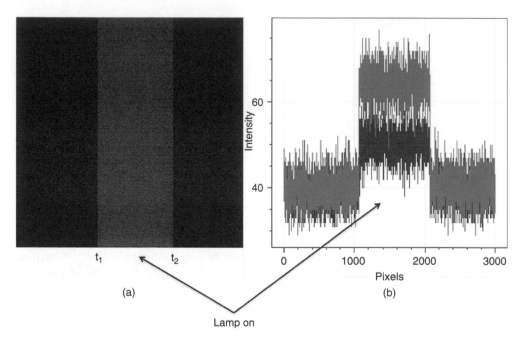

Figure 2.2   (a) Scanned image from GenePix 4000B scanner (Molecular Devices, Sunnyvale, CA, USA) without microarray slide. The image has been rotated 90° counterclockwise so that the first pixel scanned is in the bottom left corner. At t_1, a 55W incandescent table lamp was turned on approximately 15 cm away from the vents on the upper rear portion of the scanner. At time t_2 the lamp was turned off. (b) Average signal intensities for each column of pixels of image (a). When the lamp is turned on, there is an increase in the measured signals from both the red (black line) and green (grey line) channels.

2.3.8   Image Analysis and Data Extraction

The last step in the experimental process is the extraction of signals from the image created by the microarray scanner. The image constitutes the raw data for a microarray experiment. Probe level signals are extracted from the image by first applying a mask that spatially segments probe features from background and appropriately annotates these features. The placement of the mask, or grid, is usually automated for commercial arrays. However, automatic and manual placement of the grid may be hindered by poor image quality or microarray manufacturing problems. Once image segmentation is complete, the signal features and background levels are calculated. Probe level signals are calculated by subtracting the background signal from the feature signal. These calculated values are then used as the starting point for the statistical analysis and interpretation of the data.

Manual methods of extracting data from a microarray image can be subjective and may not be reproducible. Different image analysis methods and software can generate different results (Yang *et al.* 2002a). Care should be taken to make sure all image data is treated equivalently.

2.3.9   Clinical Diagnosis

Many microarray experiments or studies are done to gain a better understanding of the biology of a particular physiological state or to correlate gene expression patterns with clinical diagnosis or outcome. Currently, the gold standard for clinical diagnosis for the majority of surgically resected specimens is histopathology. Histopathological diagnosis involves the scoring of tissues and specimens under a microscope by trained personnel. The diagnosis or scoring of specimens is subjective and can be different based on who is scoring a particular specimen (Furness *et al.* 2003). Thus, a clinical bias can be introduced into an experiment if care is not taken to control for this systematic error.

A number of methods can be applied to help minimize the effect of variability in subjective clinical diagnosis. A single individual or a single lab can be used for clinical diagnosis of specimens. This will lessen any variability between or within institutions. A voting system can also be used to grade specimens. Either multiple pathologists can view and score each specimen or software can be used to score specimens (Daskalakis *et al.* 2008). The use of a voting system or software to diagnose clinical samples may be less subjective but can also introduce additional systemic biases that should be considered before adopting such an approach.

2.3.10   Interpretation of the Data

Interpreting the data generated during a microarray experiment can involve a number of steps. Typically, an algorithm is developed to classify one physiological state from another. The algorithm is first optimized using a subset of the data, called the training data set. The algorithm is then tested on the remaining data, called the test data set, to validate the algorithm. Since the algorithm is specifically designed to classify a selected data set, there is the potential to systematically bias results due to this selection (Ambroise and McLachlan 2002). The potential bias can be accessed by either cross-validation or the bootstrap method. Additionally, randomization of how data sets are picked and training is done can reduce selection biases.

2.4   Conclusions

The DNA microarray has become one of the most useful and most often used tools in the scientist's repertoire of techniques for studying the genome. However, there are many steps and procedures involved in conducting a microarray experiment and each step can introduce systematic biases in generated data. Therefore, it is important that the experimental design of each study is planned accordingly to minimize or eliminate these biases. Additionally, each experimental step should be optimized in such a way that the inherent variability in these steps does not mask the information sought.

3

Experimental Design

Peter Grass

Abstract

The task of experimental design is to design a study in as economical a way as possible and simultaneously to optimize the information content of the experimental data. Design concepts include the different types of variation, including experimental error, biological variation and systematic error, called bias. A batch effect refers to systematic errors due to technical reasons. The conclusion drawn from the study results should be valid for an entire population, but a study is conducted in a sample with a limited number of experimental units (patients) where several observational units (measurements) are taken from each under the same experimental condition. Measures to increase precision and accuracy of an experiment include randomization, blocking, and replication. An increased sample size leads to a greater statistical power. Blinding, crossing, choice of controls, symmetry and simplicity of design are further means to increase precision.

3.1 Introduction

The term 'batch effect' has its origins in statistical process control, and it refers to systematic differences of a quality parameter between different production batches. A batch effect becomes significant if the average difference between batches is much larger than the within-batch random variation. In the context of microarray studies, a batch effect may be due to many more causes than just the production process of the microarray. It may encompass the entire process of tissue sampling, mRNA extraction, hybridization, scanner settings, etc.; systematic errors can occur at each of these processing steps. It is the main objective of experimental design to cope with possible sources of variation and take them into consideration in terms of appropriate design concepts, such as randomization, blocking, blinding and others.

A designed experiment is essential to prove cause and effect. By systematically varying the explanatory variables or experimental conditions we are trying to establish a cause–effect relationship. Verification of this necessitates the collection of observations,

Batch Effects and Noise in Microarray Experiments: Sources and Solutions edited by A. Scherer
© 2009 John Wiley & Sons, Ltd

and the design of the experiment is essentially the pattern of the observations to be collected.

A further purpose of the theory of the design of experiments is to ensure that the experimenter obtains data relevant to his hypothesis in as efficient a way as possible; it strives for optimization in terms of time, resources, information content and degree of relevance. There are more and less efficient ways of using resources, and in the case of clinical and animal studies the principles of optimal study design should also take ethical considerations into account.

The wide range of microarray studies encompasses the whole variety of study designs that are known from preclinical and clinical research and other life sciences. Initially, there are studies with an exploratory intention in order to generate hypotheses about, for instance, the mode of action of a drug or the biological pathway being impaired by a specific disease. These exploratory studies are followed by confirmatory studies in order to provide evidence for or reject these hypotheses and to obtain some level of confidence. For instance, a genomic biomarker signature of patients responding or not responding to an investigational compound could be derived from blood samples collected from patients participating in a classical clinical trial. Usually the sample size of such trials is limited and the patient population is restricted by stringent inclusion and exclusion criteria. The design of the study is sub-optimal for the identification of biomarkers because biomarker identification is not the main objective of the trial. The level of confidence in such a genomic signature would be low due to these limitations. Consequently, it is necessary to validate the biomarker signature in one or more subsequent studies and to obtain reliable information about its predictive performance in terms of selectivity, specificity, and generalizability. This approach is well established and in principle not different from the development of other diagnostic tools, such as diagnostic imaging techniques or laboratory tests.

There is, however, one fundamental difference between genomic studies and classical studies. Genomic, proteomic, and metabolomic studies provide usually multiple endpoints, which can be up to 50 000 in case of microarrays or more than 100 different multiplexed assays, in contrast to the single endpoint of a classical study. Multiple endpoints need to be taken into account when planning a study because the sample size of the study has to be adjusted accordingly.

In the following, the fundamental design concepts will be introduced, followed by measures to increase precision and accuracy of experimental studies, including randomization, blocking, and replication with reference to sample size calculation. There are many further measures that may help to optimize the study design such as blinding, appropriate choice of control groups, and deviation from baseline measurements. Finally, possible biases that are typical of genomic and proteomic studies are presented.

3.2   Principles of Experimental Design

3.2.1   Definitions

Several statistical design concepts need to be introduced before we can apply the principles of experimental design to microarray studies. These design concepts are generic and hold true for any experimental study, be it a cell-based assay experiment, a toxicology study, a safety and tolerability study in healthy subjects, or typical clinical phase III trial. The design of experiments is a discipline coping with the systematic evaluation of all possible sources of variation – planned and unplanned – that might interfere with the conduct of an experimental study. The ultimate goal of the design of experiments is to increase the power of statistical hypothesis testing by accounting for these sources of variation at the planning level of the experiment. A well-powered experiment reduces the rate of false positive and false negative decisions, and consequently decreases the risk of drawing wrong conclusions.

3.2.2   Technical Variation

In natural sciences, performing an experiment means observing an effect under controlled conditions, which usually involves the measurement of a given signal or the assessment of the frequency of a given event or trait being present or absent. However, each measurement process is inherently imperfect, and the measured value usually deviates from the true value by some amount, which is the experimental or measurement error. This type of error may be due to technical variations such as temperature fluctuations, small inadvertent impurities of the reagents needed for an assay, or small differences in the manual handling of the instrument. Small inaccuracies of the treatment conditions such as dosing may also contribute to the experimental error. Microarrays produce a signal per probe set or spot which is relative to the abundance of the respective mRNA of the biological sample. In analogy to other bioanalytical technologies it is crucial to accept the limitations of the measurement process and discard values below a given limit of quantification from further processing.

Any optimization of the measurement process leads to a minimization of the measurement error and ultimately to a higher precision of the measurement result. Measures to improve the precision include repeated measurement of the same sample, standardization of the measurement process, and automation.

3.2.3   Biological Variation

Biological variation is another source of variability. It is a fundamental characteristic of living beings that they exhibit biological variations between and within species which are essential for development and evolution. Clinical parameters may vary between populations (e.g. between Caucasians and Blacks) and between individuals within a population (e.g. between males and females). Even within a subject a parameter may vary over time due to circadian rhythms, according to health conditions, demographic particularities of subjects or other nutritional and environmental factors.

In clinical chemistry, normal ranges of laboratory parameters have been defined within which the measurement values may vary under normal physiological conditions. Generally, biological variation cannot be minimized by a more precise measurement, but it can be reduced by specifying it for more homogeneous groups (e.g. for males and females separately and within each gender for different age groups). Measurement errors as well as biological variation are known to be random variations and both are very often normally distributed; that is, small positive and negative random variations around an average value are more frequent than large variations. This pattern gives rise to the well-known bell-shaped Gaussian distribution curve.

3.2.4   Systematic Variation

In addition to random variation, there may be systematic deviation of the measurement value from its true value, which is called bias. Such systematic deviations may be due to flaws in the production and measurement process. Incorrect calibration of an instrument, for instance, would yield concentrations of a laboratory parameter being measured systematically either too high or too low. In microarrays, batch effects due to, for instance, different scanner settings belong also to the class of systematic variations.

However, systematic errors may also be introduced at the level of recruitment when patients or subjects are not representative of the entire population. For instance, only young healthy males are included in a study with the objective of assessing the signal of a biomarker, but conclusions about the validity of the biomarker distribution will be made for the entire population.

3.2.5   Population, Random Sample, Experimental and Observational Units

A population encompasses the entirety of subjects under a given experimental condition, say all patients suffering from Alzheimer's disease. An experimental study, however, will never be conducted on the whole population, but rather on a small subpopulation, referred to as a sample, which would include a limited number of N subjects only. It is assumed that the sample is a valid representation of the entire population if the sample elements are drawn randomly from the entire population and the sample size N is sufficiently large.

Within a population, the parameter of interest has a given distribution which may be characterized by its population mean (μ) and standard deviation (σ) in the case of a normal distribution. The mean and standard deviation as calculated from the sample data are only estimates of the true population parameters where the precision of the estimate is influenced by the sample size.

The individual elements of a sample that undergo a predetermined treatment or belong to a certain condition are called experimental units (e.g. patients, subjects, animals), while repeated measurements of a parameter in the same experimental unit are called observational units (e.g. several blood pressure readings in the same patient obtained at the same time under the same treatment condition).

3.2.6   Experimental Factors

Experimental factors, such as conditions and treatments, are an important design concept; conditions may refer, for instance, to the disease status of a patient, and treatment may refer to different medications. Notably, experimental conditions are under the control of the experimenter and are not subject to random fluctuation. Factors are usually explanatory variables and have usually two or more levels. By definition, it is impossible to establish a cause–effect relationship without the systematic variation and controlling of experimental factors.

The investigation of the influence of different conditions or treatments on a response parameter is usually the reason for conducting a study. A main objective of experimental design is to determine systematically which conditions or treatments and which levels should be chosen. Random allocation of the experimental units to the different conditions or treatments is mandatory. The choice of an appropriate control group is very important and affects the outcome of the study.

In contrast, blocking factors are conditions or treatments that are expected to influence the measurement systematically but are of no particular interest, so-called nuisance variables. Typical examples of nuisance variables in genomic studies are ethnicity and gender of patients, or the centers of a multicenter trial, but also operators of the laboratory where mRNA extraction is performed. Confounding means the 'mixing' of sources of variation in data so that their effects cannot be distinguished from each other. Confounding factors can be introduced by the experimenter, by the patient or by the environment (Wooding 1994). A general rule (Box *et al.* 2005) says 'Block what you can, randomize what you cannot'. One of the most prominent objectives of experimental design is to identify possible nuisance variables and account for them by using them as blocking factors.

3.2.7 Statistical Errors

When comparing two samples obtained from two experimental conditions, there are two main types of error that may occur and that we want to control. Type I error (α) is defined as the risk of concluding that conditions differ when, in fact, they are the same. The α level is usually set at 5%. Type II error (β) is the risk of erroneously concluding that the conditions are not significantly different when, in fact, a difference of a given size δ or greater exists. Commonly chosen values of β are between 5% and 20%. The Type II error is directly linked to the power of a test ($1 - \beta$) which implies the probability of being able to demonstrate a statistically significant difference between the samples if a true difference of the estimated size δ actually exists in a larger population. The confusion matrix of Table 3.1 shows the four possible decisions for a typical two-class problem;

Table 3.1   A confusion table compares the true conditions with the outcome of a test procedure, for instance a diagnostic device.

		Actual (true) condition	
		Present Samples are different	**Absent** Samples are not different
Test result	**Positive** reject H_0	**True Positive**	**False Positive** (Type I error)
	Negative accept H_0	**False Negative** (Type II error)	**True Negative**

we want to maximize the correct decisions by correctly identifying the true positives and true negatives, and simultaneously we want to minimize the number of false decisions, that is, the number of false positives and false negatives.

Given the inherent variability of the data and a relevant difference we wish to detect, the only way to control both types of error is to increase the sample size. A large number of false positives is caused by insufficient specificity of the test, and a large number of false negatives may be due to inadequate sensitivity. The power of a statistical test is the probability of rejecting a false null hypothesis; as statistical power increases, the probability of committing a Type II error decreases.

3.3 Measures to Increase Precision and Accuracy

Accuracy and precision are two characteristics that determine the quality of an experimental study. Precision describes how well the measurement can be reproduced and is usually reported as an average replication error. It can be determined by performing repeated measurements of the same experimental unit. Accuracy describes how close to a true population value a measurement lies. Calibration samples of known concentrations are used to estimate the accuracy of a measurement (Moreau *et al.* 2003). In many situations, however, the true population value is unknown; this is true in particular for microarray experiments due to the absence of calibration samples. Bias affects accuracy and if the bias is large, a measurement may be of high precision but of low accuracy, as depicted in Figure 3.1.

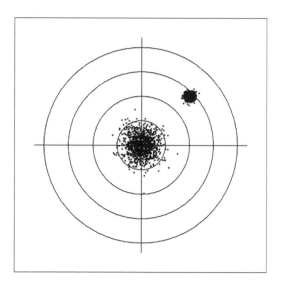

Figure 3.1 In analogy with this shooting competition target, a measurement can be highly precise (i.e. highly reproducible with little variation) but completely inaccurate because the values are systematically off center. Alternatively, a measurement may be quite accurate (i.e. the values scatter around the center) but show high variability and hence, less precision. Obviously, we strive for high precision combined with high accuracy.

In experimental design, there are three ways to increase the precision and accuracy of experimental studies: (a) refinement of experimental techniques; (b) careful selection of subjects with additional information about them that can account for a part of variation or by grouping them according to certain characteristics; and (c) enlargement of the size of the experiment by employing more replicates. Fundamental techniques applied in statistical design (Kerr 2003) include randomization, blocking and replication.

3.3.1 Randomization

A random sample is one in which each individual in a population has an equal chance of being selected. Randomization will reduce bias, and, on average, the estimates of the population parameters will be accurate (Bolton and Bon 2004). Furthermore, we assume that individuals being randomly sampled from a population are valid representatives of this population if the sample size is sufficiently large. Randomization encompasses both random selection from a population and random assignment to a treatment.

Another reason for randomization is that statistical theory is based on the idea of random sampling. In a study with random allocation, the differences between two treatment groups behave like the differences between random samples from a single population. We know how random samples are expected to behave and can compare the observations with what we would expect if the treatments were equally effective (Altman and Bland 1999).

Randomization does not eliminate the possibility that treatment and control groups will differ according to some unknown factors that affect outcome (nuisance variables). However, we can expect the random assignment of treatments to ensure some rough balancing of those factors across all treatment groups. It is important to recognize that because randomization relies on the averaging of sampling variation, in studies with small numbers of subjects it may not effectively reduce bias (Sica 2006).

3.3.2 Blocking

Blocking is another measure to increase precision and to reduce bias. In practice, subjects of similar characteristics are grouped together in blocks and randomly assigned to treatments. Common blocking variables include gender, age group, and ethnicity. In crossover trials, where the individual patient represents a block, each patient receives each of the two or more treatments in random order. Such a randomized block design is advantageous when the level of responses is very different between patients, but within-patient variability relatively small. If the subjects are properly randomized within the blocks, data analysis techniques, such as analysis of variance, are able to separate the variability due to the different blocking factors, resulting in a decreased experimental error.

3.3.3 Replication

In analogy to biological and technical variation, we can specify biological and technical replication. Biological replication is directly linked to the number of experimental units (e.g. biological samples, animals, subjects or patients) included in the study. Technical replication refers to repeated measurements of a particular parameter derived from one and the same experimental unit under the same experimental condition.

Whatever the source of the experimental error, replication of the experiment by increasing the sample size steadily decreases the error associated with the difference between the average results for two treatments/conditions. Precautions have to be taken to ensure that one treatment/condition is no more likely to be favored in any replicate than another, so that the errors affecting any treatment tend to cancel out as the number of replications is increased (Cochran and Cox 1962).

The estimate of a population mean becomes more and more precise with increasing sample size. With a given sample size N the difference δ between the two populations and the variability σ, in particular the ratio σ/δ, play a crucial role in the estimation of the population parameters. Large ratios due to higher variability and/or smaller differences yield less reliable estimates, and vice versa. Sufficient replication renders an experiment powerful enough to detect a relevant effect of size δ; large experiments have a greater chance of detecting important differences.

In the simple case of a two-class comparison, only four parameters are needed to calculate the sample size of an experimental study; they include the minimum relevant difference δ between two samples that we want to detect, the standard deviation σ of the population as a measure of variability, and the Type I and Type II errors. The difference δ should be of 'practical significance' and should make sense in the context of the experimental setting. The standard deviation σ of the parameter of interest is usually taken from already existing data – from a pilot study, from data published in literature, or by making an educated guess.

It is the ratio σ/δ that impacts the sample size at a given α and β. Intuitively, we know that a greater sample size is needed when we want to detect a smaller difference δ, or when the variability is greater. If we wish to have more confidence in our results by minimizing Type I and Type II error we would increase the sample size, too. There are formulas to calculate sample size which are specific to the different experimental designs. It is beyond the scope of this chapter to provide explicit formulas for sample size calculations under the different experimental designs and it is recommended to get advice from a statistician or use an appropriate computer program. In the context of genomic studies it is important to note that classical sample size calculations are only valid for studies with a single endpoint. Genetic, genomic, proteomic, and metabolomic studies, however, are characteristic for their multiple endpoints which need to be accounted for in sample size calculations, for instance by choosing false discovery rate (Benjamini and Hochberg 1995) or adjusting the Type I error for multiple comparisons.

In microarray studies, estimation of the standard deviation is aggravated by the fact that the number of genes with a reliable signal and the associated standard deviation are different for each of the thousands of features on a chip. In addition, they deviate from tissue to tissue, and they differ between chip technologies (Lee *et al.* 2000).

3.3.4 Further Measures to Optimize Study Design

Accuracy as well as precision can be strongly influenced by how the different treatments or conditions are allocated to the experimental units. Well-known designs of biological, toxicological, and clinical studies include parallel group design and crossover design. Many other design concepts are employed for the purpose of bias elimination and reduction of the experimental error; for instance, blinding techniques, inclusion of positive control groups, and repeated measurements. The objective is always to reduce the heterogeneity of the experimental units by careful selection of patients by specifying inclusion and exclusion criteria and checking patients' eligibility. It must be pointed out, however, that patients fulfilling the eligibility criteria for homogeneous groups are usually not representatives of all those with the disease or all those for whom the therapy is intended (i.e. the target patient population).

The art of good experimental design is to find the right balance between taking into consideration as many sources of variation as possible on the one hand and the simplicity and practicability of an experimental study on the other. Without doubt, more complex study designs have more restrictions and lack flexibility; they are difficult to implement and prone to bias. For instance, the number of patients participating in crossover studies is usually smaller in comparison to parallel studies, but the impact of patients dropping out from the study is much higher.

Another important aspect is related to the symmetry of study designs. Although most data analysis methods are able to cope (to a limited extent) with unbalanced data and missing values, a certain symmetry should be strived for – equal numbers of patients per treatment group, equal numbers of visits per patient, balanced order of administration, and equal number of replicates per patient. Symmetry of design increases the power of the analysis and – even more important – makes the conduct of the study less prone to error and facilitates the quality control of the data.

Optimal study design is concerned with maximization of the information content of the study data. In a placebo controlled trial, the relative deviation of a clinical parameter under a given treatment condition from its baseline value is more informative than comparing the mean parameter of a placebo treated patient group with that of a different group of actively treated patients. Note that deviation from baseline in the absence of a placebo control group may be due to changes in environmental conditions during the intervening time period; therefore, the inclusion of a placebo group is strongly recommended. Classical pharmacokinetic/pharmacodynamic (PK/PD) studies are very often designed as placebo controlled crossover studies. Their objective is to assess the time course of the drug concentration in blood and tissue, and simultaneously to measure the pharmacological effects that might be elicited by the investigational compound or placebo in the same patient. Indisputably, such related concentration–effect time profiles provide more information than a single reading of a response parameter at a predetermined visit. However, in many situations the design is too complex for a clinical routine setting, and another caveat is a possible carry-over effect if the wash-out phase between treatments is not long enough.

The majority of studies are designed as comparative trials, and the appropriate choice of the control treatment or condition is crucial. Several types of comparison are possible: between a patient group being treated with an active compound and another patient group being treated with placebo; or between a patient group being treated with a developmental compound and another patient group being treated with the current clinical gold standard as a positive control. Even the simultaneous inclusion of placebo treatment, positive control and negative control may make sense in certain situations. Not only the choice of the control treatment is important but also the choice of the control population should not be neglected. It makes a difference whether the control group consists of age-matched healthy volunteers or age-matched co-morbid patients.

Blinding, sometimes called masking, is another measure to avoid bias. Human behaviour is influenced by what we know and what we believe. In research there is a particular risk of expectations influencing findings, most obviously when there is some subjectivity in the assessment, leading to biased results. Blinding is used to try to eliminate such bias. Blinding is particularly important when the response criteria are subjective as it is in the assessment of pain. Double-blinding usually refers to keeping patients, those involved with their management such as nurses and study monitors, and those collecting and analyzing clinical data unaware of the assigned treatment, so that they should not be influenced by that knowledge (Day and Altman 2000). It is important that the treatments look similar: same taste and smell, same colour, same number of capsules to be taken, same route of administration which is called the double-dummy method. Blinding in the assessment of the performance of a diagnostic test is crucial: those who do the assessment must be blinded; also when testing the reproducibility of a diagnostic test the observer must be unaware of the previous measurements. If not blinded, clinical trials tend to show larger treatment effects and diagnostic test performance is overestimated.

3.4 Systematic Errors in Microarray Studies

Two kinds of bias may be of particular importance for microarray studies. The first kind refers to the selection of the experimental units (i.e. which subjects are to be recruited to participate in a study or which cell line will be employed for the experiment), and the second kind of bias may be introduced at the measurement level.

3.4.1 Selection Bias

Again, the process of selecting patients should ensure that they are representative of the target population. Inclusion and exclusion criteria should not be too stringent because they may limit the generalization of the study findings. Selection bias in biomarker development comes in many forms and they usually lead to an overestimate of the sensitivity of the assay. Examples of selection bias include the exclusion of patients with mild disease who are usually difficult to diagnose. The limitation of the patient population to only those who are volunteering to participate in the study might introduce the bias that health-conscious patients are overrepresented. Patients dropping out of a study may be different from patients remaining in the study; consequently, the exclusion of those patients from the analysis will introduce a selection bias. The choice of a control group may also introduce bias; for instance, young, disease-free subjects versus nondiseased but age-matched patients (Sica 2006; Whiting *et al.* 2004).

3.4.2 Observational Bias

Observational biases may be caused by differences in methods by which information is collected and data are obtained or due to the fact that patients may under- or over-report their medical history or personal habits if they are aware of their diagnosis. Bias due to different methods being applied to different experimental groups occurs quite frequently in clinical studies. For instance, a reference test is only applied in the case of a positive study test result. The consequence is that more sick subjects than healthy subjects undergo the reference test. Patients with negative test results and those with benign-appearing lesions do not typically undergo an invasive reference test, such as a kidney biopsy. On the other hand, bias may be introduced because patients with positive test results undergo more intensive follow-up investigations and therefore, missing data are present nonrandomly for disease-free subjects. Another issue is represented by the fact that the reference standard against which a new genomic biomarker is to be compared is not 100% accurate or a surrogate reference test is used instead of the clinical endpoint; for instance, coronary angiography for the diagnosis of coronary artery disease.

Between-group differences in reporting may happen because patients with positive diagnostic results may better recall or even exaggerate their presenting symptoms. When an unblinded investigator is the interviewer and also involved in the interpretation of the results he may request details of medical history more rigorously in the case of positive test results.

3.4.3 Bias at Specimen/Tissue Collection

There are several sources of variation that are specific to genomic and proteomic experiments. The process of tissue sampling should be randomized across conditions in order to ensure uniform handling of the tissue specimen. In a toxicology study, vehicle treated animals may all be sacrificed first, followed by the low, medium and high dose group animals sequentially in order to avoid drug contamination of histopathological tissue samples. However, due to considerable time differences between last dosing and tissue harvest the tissue exposure to the drug may be different, and in consequence, mRNA response may change due to this systematic error in sample collection. Another sample collection bias refers to the time of food intake in relation to specimen collection which should be blocked for or, if blocking is impossible, should be randomized. The sample collection scheme should also take circadian and seasonal changes into account (Rudic *et al.* 2005) and stress exposure of the patients and animals should be balanced across treatments. The brain is a complex organ, with a heterogeneous distribution of distinct subpopulations of cells, intricate signaling and regulatory circuits, and exquisite lifelong sensitivity to environmental variation. These factors result in high levels of inter-individual variability of gene expression (Karssen *et al.* 2006). Due to a difficult sampling situation in pathology, tissue contaminations may occur; for instance, rat duodenum samples are often contaminated by pancreas tissue. The tissue sample volume is related to the amount of mRNA that can be extracted and should therefore be balanced across conditions. In clinical trials, co-medication can considerably influence the gene expression pattern of the patients. The way specimens are stored may also impact the outcome of the experiment; nuisance variables include storage duration and temperature, and whether tissue samples are conserved in paraffin blocks or as snap frozen samples. Another source of variation is how the harvesting of specimens from heterogeneous tissues such as brain or kidney is done.

3.4.4   Bias at mRNA Extraction and Hybridization

The mRNA extraction protocol should be standardized and the samples should be randomized across operators. Batch effects may be introduced by age and impurities of reagents. It is a well-known fact that microarray chips vary between production batches and it is strongly recommended to use one batch of chips for one study. If this is not possible the samples should be blocked by batches. It is good experimental practice to record temperature, experimenter, and date/time of hybridization (Coombes *et al.* 2002; Li, Gu and Balink 2002; Zakharkin *et al.* 2005). Systematic variation can be due to the use of different scanners or scanner settings, or different photo-multipliers and gains. Different spot-finding software and different grid alignments may also introduce bias.

Needless to say, that the entire technical process should be standardized as much as possible and tissue sample processing should be as homogeneous as possible. If for any reason this is not possible a proper blocking or randomization of the sample processing should be taken into consideration. Obviously, technical variation of the measurement process can be controlled to a certain extent.

3.5   Conclusion

The basic concepts of experimental design are also applicable to microarray studies; randomization, blocking and replication are the most important measures to improve the accuracy and precision of the experimental outcome. Typical sources of random as well as systematic variation known from classical preclinical and clinical studies could possibly be identified also in genomic studies. There are, however, additional sources of bias that are directly related to the new technologies, and this requires the extension of the design concepts also to the tissue sampling and measurement level. From an experimental design point of view there are some fundamental issues that are not yet completely solved and need further investigation.

Genomic and proteomic studies produce information of unparalleled wealth. The high sensitivity of these technologies, in combination with the simultaneous interrogation of a large number of features (genes or proteins), is without doubt a great achievement for genome-wide screening purposes. The down-side of this is that nuisance variables which are negligible in the clinical routine may gain importance on the molecular level; factors such as co-medication or co-diseases or personal habits such as cigarette smoking may interfere with the molecular signature of a disease or a drug treatment under investigation.

When designing a genomic study one should keep in mind that gene expression is a highly dynamic phenomenon that develops over time and may be subjected to feedback and feed-forward mechanisms (Yang and Speed 2002). Therefore, the time point at which tissue specimens are collected in relation to a stimulus such as the administration of an investigational drug is important. Consequently, the gene signature is very much dependent on the time point of sampling.

Another open issue is the lack of microarray validation in a form as we know it from other bioanalytical assays, although promising attempts in this direction have been made (van Bakel and Holstege 2004). Microarray technologies, whether single- or dual-channel microarray designs, always produce relative gene expression values. At the moment it is impossible to obtain an absolute measure of the concentration or the amount of a specific mRNA segment. Validated assays would employ calibration samples for the calculation of a calibration curve and to determine upper and lower limits of quantification and detections, respectively. Quality control samples of known content could be randomly sprinkled into the set of unknown samples in order to provide in-process quality control.

As long as we do not know the true expression values of the genes it is extremely difficult to identify measurement bias that may be introduced at the mRNA extraction and hybridization level. Real-time polymerase chain reaction confirmation will always be limited to a subset of genes and cannot replace the validation of an entire microarray (Wang *et al.* 2006). Furthermore, it cannot provide information about the selectivity and precision of the microarray itself. From an experimental design point of view, dual-channel microarrays do not really provide advantages over single-channel microarrays. Quite the reverse is true; due to possible dye effects, special designs of the hybridization need to be used which include dye swapping and dye balancing in order to take these systematic errors into account (Kerr and Churchill 2001a, 2001b; Churchill 2002; Wit and McLure 2004).

A related open issue is the poor reproducibility of genomic results across studies. Even the same tissue specimen being analyzed by different laboratories may produce different genomic signatures even when using the same microarray technology. The reason is only partly due to variation in mRNA extraction protocols, hybridization process and scanning. In particular, in the field of biomarker research it had to be realized that many peer-reviewed studies publishing genomic biomarker signatures based on statistically significant differences are not reproducible, and the conclusions drawn are overly optimistic (Halloran *et al.* 2006). Validation studies need to be conducted by several completely independent teams with appropriate sample size. Small sample sizes might actually hinder the identification of truly important genes (Ioannidis 2005). In contrast to the current standard, thousands of samples are needed to generate a robust gene/protein signature for predicting outcome of disease (Ein-Dor *et al.* 2006). If the identification and validation of gene signatures are the objective, the approach of sample size calculation needs rigorous reconsideration.

4

Batches and Blocks, Sample Pools and Subsamples in the Design and Analysis of Gene Expression Studies

Naomi Altman

Abstract

Microarray experiments may have batch effects due to samples being collected or processed at different times or by different labs or investigators. Linear mixed models which incorporate the known sources of variation in the experiment can be used as design and analysis tools to assess and minimize the impact of batch effects and to determine the effects of design decisions such as sample pooling or splitting.

4.1 Introduction

Consider a microarray experiment to compare gene expression in several tissues in a complex organism such as the fruit fly (*Drosophila melanogaster*) with the ultimate goal of understanding biological issues such as the genes involved in organ development. Biologically meaningful findings should be reproducible using different measurement instruments by different research groups.

Each fly yields only a small sample from each tissue, each of which yields only a small quantity of RNA so that sample pooling may be required. Differences due to extraneous factors such as age or stress need to be minimized. If one person does the experiment, then tissue collection, RNA extraction and hybridization may need to be spread over a long period of time. Samples and reagents will need to be stored, or fresh batches need to be used. Alternatively, several people could be involved, possibly in different labs, which might shorten the timeline but could introduce variability in fly rearing protocols or in

Batch Effects and Noise in Microarray Experiments: Sources and Solutions edited by A. Scherer
© 2009 John Wiley & Sons, Ltd

experimental techniques. All of these factors can introduce variability or create bias in measuring gene expression.

A typical comparative microarray experiment such as the one outlined above includes several treatments, conditions which are to be compared, such as different tissues. Other factors producing change in estimated gene expression that is consistent across two or

| Growing the organism → Tissue sampling → RNA processing → Hybridization → Data extraction |

Figure 4.1   The main steps in performing a microarray experiment.

Table 4.1   The stages of a microarray experiment and some of the batch effects introduced at each stage.

Stage	Batch effect
Growing the organism	geographical location
	time of year
	location of 'housing' (green house, Petri dish ...)
	batch
	parent/litter
	rearing protocol (feed, water, temperature) ...
	personnel effects*
Tissue sampling	age of plant/animal
	time of day
	developmental stage of tissue
	wounding effects
	tissue boundary definition
	environmental conditions**
	personnel effects*
RNA processing	batches of reagents
	environmental conditions**
	personnel effects*
	equipment effects
Microarray hybridization	microarray platform
	batches of reagents
	environmental conditions**
	personnel effects*
	equipment effects
Data extraction	array handling (storage, fixing, washing)
	scanner effects
	personnel effects*
	post-processing effects†

*Personnel effects are any effects introduced by having a different person handle this stage of the experiment.
**Environmental conditions are effects introduced by local conditions at the time the experiment is done, such as lighting, temperature, and air quality.
†Post-processing effects include factors such as data 'cleaning' and normalization.

more samples are considered to be batch effects. Batch effects may create bias or variance in the gene expression measure, depending on whether they are considered to be reproducible effects or simply sources of noise.

Batch effects provide opportunities to improve the precision of the experiment through appropriate design and analysis, but might also invalidate the conclusions of the experiment if not appropriately accounted for. In the fruit fly experiment, batches of flies reared together share a common environment, and hence observed gene expression differences in tissues within the batch are unlikely to be due to environmental differences. On the other hand, the biological variation in flies reared together may not be representative of the variation in different batches, so that conclusions drawn from a single batch may not be reproducible.

4.1.1 Batch Effects

Figure 4.1 is an idealized timeline of a microarray study. At each time, batch effects can be introduced, either by design or inadvertently. Sources of variation are described and discussed in Chapters 2, 3, and in Table 4.1. In addition, if the experiment is repeated at another time or site, other batch effects may be recognized that were not accounted for in the original experiment.

Three important tools for controlling batch effects are a stringent experimental protocol, statistical experimental design, and statistical adjustment. Batch effects not controlled by the experimental protocol need to be accounted for in the experimental design and analysis. In Section 4.2 we develop a statistical model for the responses in a microarray study and then show how this model assists in understanding batch effects. In Section 4.3 we use the statistical model to determine the efficacy of some common experimental designs. In Section 4.4 we consider the effect of microarray normalization and statistical standardization on batch effects. In Section 4.5 we look at the use of pooled samples and subsamples or technical replicates. Section 4.6 outlines some of the reasons for doing pilot experiments. Section 4.7 is a brief summary.

4.2 A Statistical Linear Mixed Effects Model for Microarray Experiments

We denote the response for a single channel of a single microarray probe as Y_{ijkr}. The subscripts i, j, k, r are labels that identify the gene, treatment, batch, and microarray, respectively. Y_{ijkr} is usually a measure of hybridization intensity such as a scanner reading, and may be recorded on a natural scale that is (in principle) directly proportional to the number of transcripts in the sample or on a transformed scale which may be preferable for statistical analysis or interpretation. Often analysis is done on the logarithmic scale, base 2, for which fold changes are expressed as differences. The logarithmic scale also equalizes the variability of low and high expressing genes so they are more comparable than on the natural scale.

A linear model is a way of expressing the response as a sum of effects associated with treatment or batch component. The effects may be fixed (systematic) or random, in which case we expect that other randomly selected levels might be used in future experiments.

Fixed effects contribute to the systematic effects in the study; random effects contribute to variability and correlation.

For example, in an experiment with one treatment factor and one batch factor, the model is usually written as

$$Y_{ijkr} = \mu_i + T_{ij} + B_{ik} + (TB)_{ijk} + A_{ir(jk)} + e_{ijkr}, \qquad (4.1)$$

which is a very simplified version of equation (1) in Altman (2005). Here μ_i is the population mean response over all treatments and batches. T_{ij} is the deviation of the mean response of gene i under treatment j compared to its overall mean response averaged over the batches. B_{ik} is the deviation of the mean response of gene i in batch k compared to its overall mean response averaged over treatments. $(TB)_{ijk}$ is the the interaction between the treatment and batch effects: the deviation of the mean response of gene i under treatment j in batch k from $\mu_i + T_{ij} + B_{ik}$. The left-hand panel of Figure 4.2 illustrates a response with a treatment effect and a batch effect but no interaction, while the right-hand panel illustrates a response for which the treatment and batch effects interact.

$A_{ir(jk)}$ is the effect of array r. The notation $ir(jk)$ indicates that array is nested within the treatment and batch effect – that is, each array includes a unique batch and treatment combination (or two batches and two treatments for two-channel arrays). e_{ijkr} is random error which is specific to the measurement of probe i on array r.

The designation of factors as fixed or random is conceptual and depends on the type of inferences to be made. Quantitative factors such as temperature for which inferences such as linear or nonlinear effects or optimal levels are to be estimated are considered fixed. Qualitative factors such as genotype may be considered fixed if inference is made only for the levels observed in the study, or if there are a very small number of levels. If the levels represent a larger population then the qualitative factor is considered random. Batch effects may be fixed if they are under experimental control (such as grouping animals by

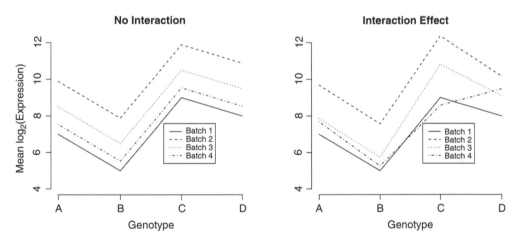

Figure 4.2   An idealized outcome from model (4.1) when there is no interaction between the treatment and batch effects (left frame) and when there is interaction (right panel). The lines connect the means from the same batch.

weight) but are usually considered to be random if they differ among experiments. Fixed batch effects may bias the experimental results. Random batch effects increase variance between batches and induce correlation within batch.

The definition of effects as deviations implies certain constraints which assist in interpretation. Fixed effects sum to zero and random effects have mean zero. An interaction is considered random if any of its components are random. So, in the usual case that the treatment is fixed and the blocks are random, we have

$$\sum_{j=1}^{r} T_{ij} = 0, \sum_{j=1}^{r} (BT)_{ijk} = 0, E(B_{ik}) = E\left((BT)_{ijk}\right) = E(e_{ijkr}) = 0.$$

4.2.1 Using the Linear Model for Design

There are two principles for using the model to help understand batch effects: the subtraction principle and the averaging principle.

According to the subtraction principle, if two measurements share the same level of an effect, their difference is free of the effect. So, for example, if a batch effect is due to the date of the hybridization k, and the two samples are hybridized to the same array r, then the difference in measurements for the same gene i on a single array is

$$Y_{ijkr} - Y_{ickr} = \mu_i + T_{ij} + B_{ik} + (TB)_{ijk} + A_{ir(k)} + e_{ijkr} - (\mu_i + T_{ic} + B_{ik} + (TB)_{ick}$$

$$+ A_{ir(k)} + e_{ickr})$$

$$= [T_{ij} - T_{ic}] + [(TB)_{ijk} - (TB)_{ick}] + [e_{ijkr} - e_{ickr}]. \tag{4.2}$$

The difference is free of effects of the overall mean, the main effect of batch and the array effect. This principle underlies the common use of the difference between the channels (i.e. the logarithm of the fold change) on two channel microarrays.

The averaging principle says that if a and b are numbers, and X and Y are independent random effects, the variance of $aX + bY$ is $a^2\text{Var}(X) + b^2\text{Var}(Y)$. A consequence of this is that the average of n independent random effects is the variance of the effect divided by n. Another consequence is that the variance of the difference of independent random effects ($a = 1$ and $b = -1$) is the sum of the variances. So, in equation (4.2), the variance of the difference of the two observations is twice the batch by treatment variance and twice the random error variance.

4.2.2 Examples of Design Guided by the Linear Model

The subtraction and averaging principles can be used with the model to predict the behavior of the responses in an experiment and to design the experiment to reduce batch effects and hence improve precision of the comparisons of interest. Some examples are below.

Example 4.2.1: Complete replication of the experiment within a batch
Suppose we plan an experiment with a known batch effect and we can arrange for one sample from each of the t treatments to be observed in each of b batches. For example,

the effect could be due to a batch of reagents which is sufficient for t samples, or a lab technician who can process t samples in a single day. Then, averaging all of the samples (i.e. one copy of each treatment in each block), we obtain

$$\overline{Y}_{i...} = \frac{1}{bt} \sum_{k=1}^{b} \sum_{j=1}^{t} Y_{ijkr} = \mu_i + \overline{B}_{i.} + \overline{A}_{i..} + \overline{e}_{i...}$$

If the batch effects are fixed, then $\overline{B}_{i.}$ is zero; if random, then the variance of $\overline{B}_{i.}$ is $Var(B)/b$. The variance of $\overline{A}_{i..}$ and the variance of $\overline{e}_{i...}$ are both reduced by a factor of $1/tb$ compared to the variance of a single observation. So the mean over all the samples in the experiment is a slightly noisy estimate of μ_i, the average expression level of gene i. Similarly, by averaging over all the samples in the kth batch and jth treatment respectively, we can obtain slightly noisy estimates of the batch effect k and treatment j effects.

Usually, we are most interested in estimating the difference in the means of treatments j and c, that is, $T_{ij} - T_{ic}$. The difference between treatment averages is

$$\overline{Y}_{ij..} - \overline{Y}_{ic..} = [T_{ij} - T_{ic}] + [(\overline{TB})_{ij.} - (\overline{TB})_{ic.}] + [\overline{A}_{ij.} - \overline{A}_{ic.}] + [\overline{e}_{ij..} - \overline{e}_{ic..}] \quad (4.3)$$

Notice that estimate is free of any block effect, and the variance of the estimate is reduced by a factor of $1/b$ compared to the variance of difference between the measured responses in a single batch.

Example 4.2.2: Single batch for the entire experiment
Alternatively, if the entire experiment can be done within a single batch, then the difference in treatment means has no batch effect. However, there is no averaging of the batch by treatment interaction. The interaction induces a bias in estimating the differences between treatments, which cannot be determined from the experiment but which affects reproducibility if the experiment is repeated. For example, if the batch effect is due to the investigator performing the experiment, the estimate of treatment effects may be highly replicable when the experiment is repeated by this investigator. However, another investigator may not be able to reproduce the effect. If the batch by treatment interaction is small, running the experiment as a single batch is often a good strategy for controlling variability.

Example 4.2.3: Batch confounded with treatment
It often appears to be convenient to run all the replicates of each treatment in separate batches. For example, in a test of chemical exposure, we may rear all exposed animals in one batch and all the unexposed animals in another. Let us suppose all the replicates of treatment j are in batch k and all the replicates of treatment c are in batch d. Then

$$\overline{Y}_{ijk.} - \overline{Y}_{icd.} = [T_{ij} - T_{ic}] + [B_{ik} - B_{ic}] + [(TB)_{ijk} - (TB)_{icd}] + [\overline{A}_{ij.} - \overline{A}_{ic.}]$$
$$+ [\overline{e}_{ijk.} - \overline{e}_{icd.}] \quad (4.4)$$

Notice that in this design, the differences in treatment means is the sum of the difference in treatment effects, in batch main effects and in batch by treatment interaction. All the other sources of variation are reduced by the averaging principle, but no matter how many replicates are used, the treatment comparisons are contaminated by batch differences. Not only will the results fail to be replicable, they may be completely incorrect unless the batch effects are much smaller than the treatment effects.

4.3 Blocks and Batches

An experiment is said to have blocks when randomization of samples to treatments is done within sets of similar samples. Similarity may be induced by a batch effect or may be created by splitting the samples into batches based on some characteristic such as body weight or tumor size. Blocking is an efficient way to ensure balance or partial balance for batch effects.

4.3.1 Complete Block Designs

A block is said to be complete if it contains at least one replicate of every treatment. A complete block is balanced if all treatments have the same number of replicates within the block. If all the batches in an experiment are used as balanced complete blocks then the batch effect will be removed from all treatment comparisons, as in Example 4.2.1.

Considering the expression for the treatment mean from Example 4.2.1, we can see that having more blocks with one replicate per block reduces the variance of all the average error terms. If, instead, the number of replicates within a block is increased, the block by treatment interaction variance is not reduced, since the number of blocks is not increased. Unless this interaction is negligible, it is generally more effective to have more blocks with one replicate per block, rather than fewer blocks with more replication. This is the usual 'randomized complete block design' experiment.

Example 4.3.1: Designing an experiment over several sites
Suppose that an experiment with four treatments is supposed to be done with three replicates, and four labs are cooperating on the project. If each lab processes one of the treatments, then differences between treatments follow equation (4.4) and it is impossible to know if observed differences between treatment means are due to the lab effect or the treatment effect. If, instead, three of the labs process one replicate of each treatment, then differences between treatments follow equation (4.3). The differences between treatment means are estimated free of the lab main effect, and the variance due to random error and treatment by lab interaction is reduced by a factor of 3.

4.3.2 Incomplete Block Designs

It is not possible to use complete blocks if the number of treatments is larger than the number of samples available in the block. However, it is still possible to have partial balance, which eliminates some although not all of the block effect in comparisons. In a balanced incomplete block design, each pair, triplet, ... of treatments occurs in the same block equally often. Hence, part of the block mean is in common in each pair of treatment means, and therefore cancels.

Returning to our example, suppose that each of the four labs can handle at most three samples. Labeling the treatments C, D, E, and F, the best allocation of samples to labs would be in sets of three (CDE), (CDF), (CEF) and (DEF) for which each pair of treatments is in the same block twice. Then, for example, the estimated mean difference between treatments C and D is:

$$\overline{Y}_{iC..} - \overline{Y}_{iD..} = [T_{iC} - T_{iD}] + \left[\frac{1}{3}(B_{i1} + B_{i2} + B_{i3}) - \frac{1}{3}(B_{i1} + B_{i2} + B_{i4})\right]$$

$$+ \left[\frac{1}{3}\{(TB)_{iC1} + (TB)_{iC2} + (TB)_{iC3}\} - \frac{1}{3}\{(TB)_{iD1} + (TB)_{iD2}\right.$$

$$\left. + (TB)_{iD4}\}\right] + \overline{A}_{iC..} - \overline{A}_{iD..} + \overline{e}_{iA..} - \overline{e}_{iB..}$$

$$= [T_{iC} - T_{iD}] + \left[\frac{1}{3}(B_{i3} - B_{i4})\right] + [(\overline{TB})_{iC.} - (\overline{TB})_{iD.}] + \overline{A}_{iC} - \overline{A}_{iD..}$$

$$+ \overline{e}_{iC..} - \overline{e}_{iD..}$$

so that much, although not all, of the block effect is missing from each comparison. The variance of $(B_{ij} - B_{ik})/3$ is one ninth of the variance of $(B_{ij} - B_{ik})$ and the variance of all other average effect differences is 1/3 times the variance of the differences between single observations.

Balanced incomplete block designs exist only for certain combinations of replicates, samples per batch and numbers of treatments. These can be extended to partially balanced incomplete block design in which some pairs are together a times and others b times and where a and b depend on the design. In a partially balanced design, some comparisons are more precise than others but there is still partial cancelation. The popular loop design (Kerr and Churchill 2001a, 2001b), is a clever application of a balanced complete block design for the red and green label effect, with each used for one sample on each array, and a partially balanced incomplete block design for the treatments, which occur together either 1 or 0 times on each array. Computations in Altman and Hua (2006) show that some comparisons have smaller variance than others in this design. However, the design is very efficient in removing a number of effects, including the label effect, from treatment comparisons.

Whenever possible, batches with large effects should be used as blocks, and the experiment should use balanced complete blocks, balanced incomplete blocks and partially balanced incomplete blocks in order of preference. Kuehl (2000) and Montgomery (2009) have more discussion of these designs.

4.3.3 Multiple Batch Effects

Most microarray experiments have multiple batch effects. The linear model can assist in the design of experiments with multiple batch effects by predicting their effects on the estimates of treatment comparisons.

For example, suppose there are two batch factors B and C, which may interact with each other and with the treatment factor, all of which have the same number of levels, R. Then the model is

$$Y_{ijkr} = \mu_i + T_{ij} + B_{ik} + (TB)_{ijk} + C_{is} + (TC)_{ijs} + (BC)_{iks} + (TBC)_{ijks}$$

$$+ A_{ijksr} + e_{ijksr}$$

One design choice is to completely confound B and C so that the blocks are $(BC)_{kk}$ with one replicate of each treatment in the block; another design with the same number of samples is a Latin square, which is a randomized complete block design in both B and C, with one sample in each B, C combination. Both designs use $R \times R$ samples. In the former choice, the B and C effects are completely confounded, that is, we can assume $k = s$. In this case we find that

$$\overline{Y}_{ij..} - \overline{Y}_{ic..} = [T_{ij} - T_{ic}] + \left[\overline{(TB)}_{ij.} - \overline{(TB)}_{ic.}\right] + \left[\overline{(TC)}_{ij.} - \overline{(TC)}_{ic.}\right]$$

$$+ \left[\overline{(TBC)}_{ij.} - \overline{(TBC)}_{ic.}\right] + \left[\overline{A}_{ij..} - \overline{A}_{ic..}\right] + [\overline{e}_{ij..} - \overline{e}_{ic..}],$$

so that all of the batch main effects and the BC interactions cancel, although there is still batch by treatment interaction. In the second case,

$$\overline{Y}_{iA..} - \overline{Y}_{iB..} = [T_{iA} - T_{iB}] + \left[\overline{(TB)}_{iA.} - \overline{(TB)}_{iB.}\right] + \left[\overline{(TC)}_{iA.} - \overline{(TC)}_{iB.}\right] + \left[\overline{(TBC)}_{iA.}\right.$$

$$\left. - \overline{(TBC)}_{iB.}\right] + \left[\overline{(BC)}_{iA.} - \overline{(BC)}_{iB.}\right] + [\overline{A}_{ij..} - \overline{A}_{ic..}] + \overline{e}_{iA..} - \overline{e}_{iB..}$$

so that only the B and C main effects cancel. The dependence of the mean BC effects on the treatment level is due to the fact that only certain TBC combinations are applied to the same experimental unit. We can see immediately that the completely confounded design will be less variable. Hence, it will be preferred when it can be used.

4.4 Reducing Batch Effects by Normalization and Statistical Adjustment

Normalization uses the responses of all or some of the probes on the array to adjust for the array effect. The objective is to remove effects due to technical variation while retaining biologically meaningful gene expression differences. Because the array is nested within treatment and batch (for one-channel arrays) or within pairs of treatments and batches (for two-channel arrays) some of the array effect is due to batch effects; hence normalization partially controls for batch effects.

Normalization methods typically consist of two main steps, background correction and statistical adjustment. For arrays that use multiple probes per gene, there is an additional step of combining the information from all the probes to create an expression summary for the gene.

Figure 4.3 shows a portion from the scanner image of a two-channel microarray. The brighter circular regions are the probes. Between the probes is 'empty' substrate. The

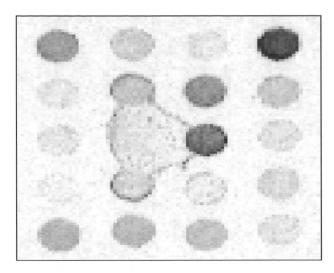

Figure 4.3   A portion of the scanner image from a two-channel microarray. The probes are printed on a grid. The spaces between the probes typically have very low intensity. However, in this view there is a local defect, elevating the background measurement locally in the center of the image.

scanner software locates the pixels on the scanned array that correspond to the probe (foreground) and the space between the probes (background) and creates a summary of the fluorescence intensity for the foreground and background of each spot. The background should be close to zero, but occasional defects or contaminants may produce signal as illustrated in the center of the figure. Often the background intensity is simply subtracted from the foreground intensity, although other algorithms are available (Edwards 2003; McGee and Chen 2006); for comparisons of methods, see Scharpf *et al.* (2007) and Ritchie *et al.* (2007).

After background correction, statistical adjustment is done to all responses. The simplest methods center all the probes by subtracting an array-specific quantity, so that the average of all the probes is the same on all the arrays in the study. The quantity removed is a statistical summary such as the mean or median expression value that is expected to be constant on all the arrays. While the summary may be computed from all the probes on the array, a subset of selected probes may also be used. The subset may be 'housekeeping genes' which are expected to express at the same level in each sample, 'negative controls' which do not hybridize with transcripts from the species in the experiment, or probes designed to hybridize to a large number of genes (Yang *et al.* 2001, 2002c; Altman 2005). Centering adjusts for the total signal from the array. It therefore removes array effects that affect every spot, such as the concentration of the RNA sample, label effects, and scanner settings. The left-hand and center panels of Figure 4.4 demonstrate the effects of centering.

There are many more sophisticated normalization methods, each with its own application and strength. For example, on two-channel microarrays, loess regression (Cleveland 1979; Yang *et al.* 2001, 2002c) is used to remove any trends in the log(difference of expression) for each probe as a function of the total expression for the probe.

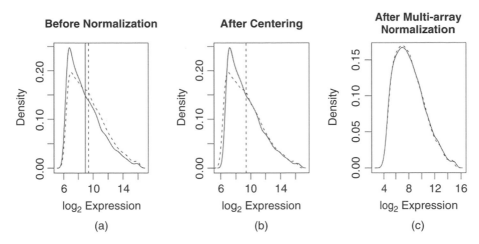

Figure 4.4 Kernel density estimate of the $\log_2$ intensity of the red channel from two microarrays in a large experiment. Left panel: the raw data for each array, with vertical lines denoting the mean of all the genes on each array. Center panel: the data after centering the arrays to have a common mean. Right panel: the normalized data after applying loess normalization to each array, quantile normalization over the entire experiment to $A = (\log_2 \text{ red} + \log_2 \text{ green})/2$ for each probe, and finally, transforming back to the single-channel normalized estimate of expression for each probe.

Centering and loess normalization are single-array methods. On each array the response for each probe is adjusted based on statistical summaries from the same array. Multi-array methods which adjust the response based on values from all the arrays in the experiment arc also used to reduce array effects. For example, a popular method is the quantile method, in which the response of the kth lowest expression level on each array is replaced by the average of the kth lowest expression level over all arrays (Bolstad *et al.* 2003; Yang and Thorne, 2003). Multi-array methods tend to be more stringent in equalizing the expression distribution across the arrays. The right-hand panel of Figure 4.4 shows the effects of multi-array normalization. Because normalization removes sample specific effects, it also removes at least some batch effects.

Normalization is generally done under the assumption that most genes are not differentially expressed. Under this condition, most of the genes should have about the same response in most samples. However, the assumption that most genes do not differentially express is not true for many microarray studies. For example, studies in the model plant system *Arabidopsis* show more than 40% of the genes differentially express between major tissues (Zhang *et al.* 2005). If a large number of genes are preferentially regulated in one direction under certain treatments, normalization can mask this effect, while inducing apparent differential expression in other genes (Figure 4.5).

4.4.1 Between and Within Batch Normalization with Multi-array Methods

Both the experimental treatments and batches affect gene expression. Multi-array normalization methods force the normalized expression values to be similar on all the arrays.

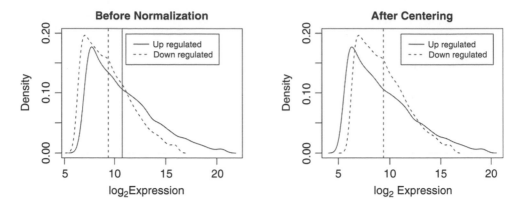

Figure 4.5   Idealized picture showing two samples with differing expression patterns. The up-regulated sample (solid line) has higher expression for most genes than the down-regulated sample (dashed line). After centering, many genes in the up-regulated sample appear to be down-regulated and vice versa. The vertical lines mark the sample means. After centering, both samples have the same mean.

Improper use of multi-array methods can introduce batch effects by making the responses more similar among arrays that are normalized together.

If multi-array normalization is used within batches or treatments, it can also induce a batch effect, because arrays normalized together will have more similar measured response than arrays normalized separately. This confounds the experimental and normalization effects. For example, Figure 4.6 shows scatterplots of the responses before and after normalization of all 10 188 probes on an Agilent® custom two-channel avocado array designed for the Floral Genome Project (Soltis *et al.* 2007). The experiment consisted of samples from eight tissues collected in a randomized complete block design with two repli-cates taken in each of two years (Chanderbali *et al.* 2009) and hybridized to two-sample Agilent arrays in two loops (data available on book companion site www.the-batch-effect-book.org/supplement). The data in the figure are the unnormalized single channel values for the four medium bud samples (Figure 4.6(a)) and multi-array normalized data for the same samples done for the whole experiment (Figure 4.6(b)) or by year (Figure 4.6(c)). Multi-array normalization was done by 'Aq' normalization across arrays, followed by loess normalization within array (Yang and Thorne 2003). The normalized data were then transformed back to single-channel values.

The scatterplots of the raw data show strong batch effects. The samples collected and processed in the same year are more tightly scattered around a trend line than samples collected and processed in different years. Furthermore, there are strong array effects that manifest as nonlinear trends in the scatterplots. The four samples displayed were all hybridized to different microarrays, with different samples in the other channel. The loess normalization was done within the two channels on each array. Nonetheless, the normalization removed the nonlinear trend on the scatterplots of samples from the same tissue. It also removed some, although not all, of the increased scatter due to the batch effect. Normalization within batch inappropriately reduced the variance within batches

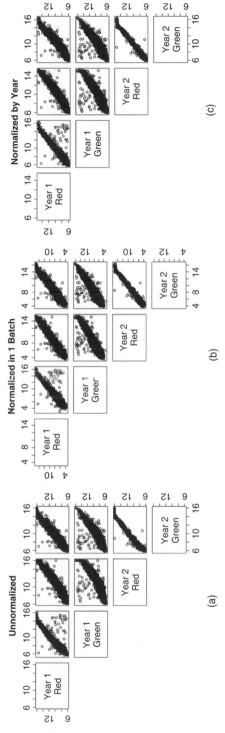

Figure 4.6 Gene expression in avocado small floral bud showing the effect of multi-array normalization on batch effects. The data come from a large loop design of eight tissues with two tissue collections and hybridization dates, one in each year of the study. Year 1: red and green are biological replicates from the first loop design experiment, done in the first year of the study. Year 2: red and green are from tissue collections and hybridization in the second year of the study. (a) Before normalization the lowest within-batch correlation is 0.964 and the lowest between-batch correlation is 0.934 and the relationship among expression levels in different samples is nonlinear. (b) After multi-array normalization of all the data the lowest within-batch correlation is 0.970 and the lowest between-batch correlation is 0.953. All pairs of normalized samples are linearly related. (c) After multi-array normalization of all the data within each year the lowest within-batch correlation is 0.970 and the lowest between-batch correlation is 0.950, and the relationship among expression levels in different years is nonlinear.

(although not a great amount in this case) but at the price of increasing variance between batches. This artificially inflates the statistical significance of the year effect and thus biases the results for tissue comparisons.

4.4.2   Statistical Adjustment

Some batch effects may be associated with a predictor variable that can be used to estimate the size of the batch effect on each gene. For example, if the air temperature in the lab on the hybridization date is linearly associated with the responses for every gene, then linear regression can be used to adjust for the air temperature effect.

We can think of a linear model for the batch effect,

$$B_{ik} = \alpha_i + \beta_i X_k + \eta_{ik}, \tag{4.5}$$

where X is the predictor variable or covariate, α_i and β_i are gene-specific intercept and slope respectively, and η_{ik} is that portion of the batch effect that cannot be predicted from the covariate. If there is a treatment by batch interaction, there is a separate intercept and slope for each treatment. Adjusting for the covariate essentially means removing the estimated trend for each gene. This method, which is commonly used in experimental situations, is called analysis of covariance or ANCOVA (Kuehl 2000; Montgomery 2009).

For example, referring to Figure 4.2, suppose that the batch effect is due to the age of the labeling kit, the ages of the kits being, in order of batch label, 6 months, 3 days, 4 months and 4 weeks respectively. Notice that the age of the kit is associated with the overall level of response in both panels. Adjusting for the age of the kit by replacing B_{ik} in equation (4.1) by the linear model in equation (4.5) will remove much of the spacing between the curves in the case without interaction, and will move the curves towards the center of the plot in the case with interaction. If the arrays were labeled several days apart, and the aging effect was large enough so that this time period could affect the outcome, X_k can be replaced by X_{kr}, the sample-specific value of the covariate. If there are sample specific effects, a sample-specific analysis of covariance will be more effective at removing the batch effect than the batch-specific analysis.

Generally estimating the batch effect as a function of a covariate is limited to simple predictors with a few parameters such as slope, because many batches are required to fit models with more parameters without overfitting. (Overfitting means that the predications are fitting the noise as well as the batch effect and hence are introducing another source of error.) Properly used, statistical adjustment can at least partially correct for batch effects. Tchetcherina (2007) discusses the use of array summary statistics such as the summaries used for centering, for a gene by gene adjustment across arrays.

Using blocking to eliminate batch effects is more effective than statistical adjustment in most cases. If the batches also include a quantitative aspect, and if the samples within the batch vary on this aspect, then blocking and statistical adjustment can be used together. For example, if four kits are used for labeling, then the experiment could be done as a randomized complete block design, blocked on kit. If the samples are not all labeled in a batch, then the actual age of the kit on the labeling date could be used as an array-specific covariate.

4.5   Sample Pooling and Sample Splitting

4.5.1   Sample Pooling

In many microarray experiments, samples from different experimental units are pooled, either before or after RNA preparation. Often pooling is done because the quantity of RNA that can be extracted from a single tissue sample is too small for hybridization.

The number of transcripts of a gene from well-mixed pooled sample is like an average of the number of transcripts from the component samples (Kendziorski *et al.* 2003, 2005; Altman 2005). Typically, pooling averages sources of variance upstream of the pooling step – that is, random effects are introduced independently prior to pooling, and then averaged. Once the samples have been pooled, they become a batch, and any further sources of variation affect the batch and hence are not averaged. Pooling can be used to increase the number of biological replicates without increasing the number of arrays. If neither the costs nor the error sources depend on the amount of material being processed, the greatest variance reduction comes from pooling as late as possible – right before hybridization. However, practical considerations may argue in favor of pooling at earlier stages.

As a simple example, consider an experiment in which samples of two tissues are taken from mice, such as two regions of the brain A and B. Tissue A can produce sufficient RNA from a single mouse for a microarray hybridization, but samples from three mice must be pooled in order to obtain sufficient RNA from tissue B. Suppose the budget allows for rearing and sacrificing 12 mice. If samples of both tissues can be obtained from the same mouse, then the mouse effect could be eliminated by using mouse as a block. Since tissue B requires a pool of three mice, which includes the average mouse effect for the mice in the pool, the appropriate block is the set of three mice. This implies that to eliminate the mouse effect, RNA samples for both tissues should be pooled samples from the same three mice. The experiment then uses four replicate pools with three mice per pool and eight hybridizations.

If each tissue must be taken from a different mouse, the mouse effect cannot be eliminated from tissue comparisons. If pools of three mice are used for each tissue, then there are only two replicate pools for each tissue and four hybridizations. If pools of three mice are used for tissue B, but individual mice are used for tissue A, then the experiment can have three replicates for each tissue and six hybridizations.

We will use equation (4.1) to analyze the design choices. Our assumption is that the random error of all samples has the same distribution, and that pooling is like averaging, which is approximately true when the pooled samples all contain about the same number of transcripts, and the response variable is proportional to the number of transcripts.

From Table 4.2, we can see that there is a large improvement in the variance of the comparison if we can take both tissues from the same mice, and use pools of three for both tissues, due to the complete elimination of the mouse effect and the use of 12 samples for each tissue. The other two designs cannot eliminate the mouse effect. Although it uses fewer hybridizations, the design with pools of three for both tissues has smaller variance than the design with pools of three only for tissue A, because the latter design has insufficient replication of tissue B.

Table 4.2   The variance of comparisons among treatments when sample pooling is needed for one of the treatments. Two-channel microarrays are assumed for the computation.

Design	Number of arrays	Mouse variance	Interaction and error variance
Pool of 3 for both tissues from the same mouse	4	0	1/6
Pool of 3 for both tissues from different mice	2	1/3	1/3
Pool of 3 for A, tissue from one mouse for B	3	4/9	4/9

4.5.2   Sample Splitting: Technical Replicates

Samples which share all the biologically induced effects are said to be subsamples or biological replicates. For example, two whole blood samples from a single mouse are technical replicates, but a sample of red blood cells and T-cells from the same mouse are not. When an experiment has only technical replication, there is a hidden batch effect – the effect of the entity that was sampled. Biological replication is always required for valid biological inferences, as it is the only way to account for effects with biological variation.

Typically, technical replicates come from splitting a sample, or taking a second sample from the same individual. Referring again to Figure 4.1, subsampling provides independent replication of all sources of variance downstream of the subsample. Hence, all sources of error downstream will have variance reduction due to the averaging principle. Technical replication can be useful if the downstream error is large and technical replication is much less expensive than biological replication. However, biological replication includes both biological and technical replication and is more effective on a per-sample basis.

A common practice which is counterindicated for biological studies is sample pooling followed by technical replication from the pool. This is illustrated in Figure 4.7. As discussed in Section 4.5.1, sample pooling averages over upstream effects, while technical

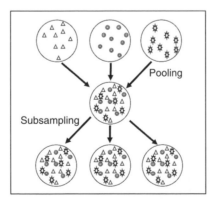

Figure 4.7   Sample pooling followed by sample splitting. In this example, three biologically distinct samples are pooled. The samples are then split into subsamples, producing three homogeneous technical replicates. The biological variance cannot be determined from the technical replicates. This design is appropriate for inference about technical errors in a microarray experiment, but not for biological inference.

replication averages over downstream effects. Thus pooling followed by technical repli-cation is no more effective for reducing variance than using smaller pools or individual samples with no technical replication, while increasing the danger that a single bad sam-ple will contaminate the pool. As well, if there is only one pooled sample per treatment, there is no measure of biological variation and hence no means of producing biologically valid inferences.

4.6   Pilot Experiments

Microarray experiments are no different than other laboratory experiments. Any practice that introduces batch effects in other experiments with the same organism is likely to produce batch effects in a microarray experiment, whether it is handling of the live specimens or storage of reagents. The best way to determine the relative sizes of effects is to run one or more small pilot experiments. Typically, if only technical batch effects are of interest, technical replication will primarily be used and several levels of the batch effects will deliberately be introduced with replication within each level.

   A pilot experiment serves many purposes. It is a training exercise for the personnel who will later run the experiments. If a new microarray platform has been developed, it is a debugging run for the array. Difficulties in the experimental processes can be worked out, from dealing with prematurely dead organisms to finding the most reliable supplier of the reagents. Planning and executing the pilot experiment forces the experimenter to think through the possible sources of error before running the main experiments, leading to improved experimental protocols to control batch effects. Finally, the pilot experiment (usually) provides preliminary data that can guide the main experiments and give insight into the size of batch effects.

4.7   Conclusions

Because of the many steps involved in performing a microarray experiment, batch effects can inadvertently be introduced. Normalization and analysis of covariance can partially adjust for batch effects. However, data processing can never fully remove batch effects, because only noisy estimates of the effects, based on an approximating model, can be removed.

   If the major batch effects are known, appropriate blocking design can be used to eliminate or reduce the main batch effects in comparisons among treatments. In contrast, poor blocking design can confound the treatment and batch effects, making it impossible to distinguish between batch and biologically relevant effects.

   Although this chapter has focused on measuring gene expression, batch effects are present in other types of microarray and high-throughput genomic studies. Often statis-tical input is not sought until after the experiment is finished and the data have been collected. Statistical design and analysis all contribute to efficient experiments, providing biologically relevant results at lower cost than experiments based solely on heuristics. To cite one of the seminal thinkers in both statistics and genetics, R.A. Fisher (1938): 'To call in a statistician after the experiment is done may be no more than asking him to perform a post-mortem examination.'

Acknowledgements

Discussions with Laura Zahn, André Chanderbali, Claude dePamphilis, Hong Ma and Jim Leebens-Mack led to many of the ideas promulgated in this chapter. Any errors are my own. Thanks also to André Chanderbali for allowing me to use the avocado data in Figures 4.4 and 4.6. Partial support for this work was provided by NSF DBI-0638595.

5

Aspects of Technical Bias

Martin Schumacher, Frank Staedtler, Wendell D Jones, and Andreas Scherer

Abstract

Variation in microarray data can result from technical and biological sources. While the extent to which technical factors contribute to this variation has been largely investigated (Bakay *et al.* 2002; Boedigheimer *et al.* 2008; Eklund and Szallasi 2008; Fare *et al.* 2003; Han *et al.* 2004; Lusa *et al.* 2007; Novak *et al.* 2002; Zakharkin *et al.* 2005), the nature and extent of the signal intensity changes with which variation manifests itself in the data has not been a major focus of research. Using several real microarray data sets with known batch effects, we analyze and describe how technical variation is translated into gene expression changes.

5.1   Introduction

When working with data from microarray experiments in which it is not possible to process all the samples (RNA extraction, labeling, hybridization, scanning, etc.) together simultaneously under identical conditions (i.e. in one batch), so-called batch effects are very often observed. Batch effects are defined as systematic differences in the gene expression intensities in samples processed in different batches which are exclusively introduced by technical factors (Zhang *et al.* 2004). The size of batch effects can be larger than the effects of biological variables which are investigated in such an experiment. Consequently, real biological effects can be distorted or even go undetected under these circumstances. The risk of false negative and/or false positive findings is clearly enhanced when batch effects are present in the data. Our focus is on large-scale microarray experiments with many samples which cannot be processed in only one batch and/or samples which were taken at times with large intervals, as is often the case in clinical trials. In such a context subsets of the samples will often be processed at different points in time (chronologically) to address specific questions before the whole experiment is completed, thus introducing

Batch Effects and Noise in Microarray Experiments: Sources and Solutions edited by A. Scherer
© 2009 John Wiley & Sons, Ltd

batch effects by definition. Another scenario which generates batch effects is the reprocessing of samples which need to be repeated for some reason. These repeated samples
are often distinctly different from the bulk of other samples processed together under
identical conditions. Unfortunately, none of the currently known normalization methods
(Stafford 2008) efficiently eliminates these technical biases. Careful experimental design
and the subsequent application of statistical methods (e.g. analysis of variance (ANOVA))
for the modeling or elimination of batch effects can effectively handle the problem (Kerr
et al. 2000b; see also various chapters in the present volume). In this contribution we are
not concerned with the detection or mathematical handling of batch effects. These aspects
are covered in other chapters of this book. Our aim is to characterize the nature and the
extent (size) of batch effects. We will attempt to assess, *inter alia*, whether batch effects
depend on the nucleotide sequence of probes, expression intensity, their characteristics
with regards to their direction, which proportion of probe sets is affected, and whether
batch effects in experiments with multiple batches are constant or variable.

5.2   Observational Studies

5.2.1   Same Protocol, Different Times of Processing

Technical bias can occur at various sample processing steps. An illustration of the deviations that one can see when one slightly changes a protocol or processes RNA from
the same source at different times, in this case the Universal Human Reference RNA
(UHRR), is provided in Figure 5.1. Hybridizations of UHRR (Stratagene, Inc.) to a human

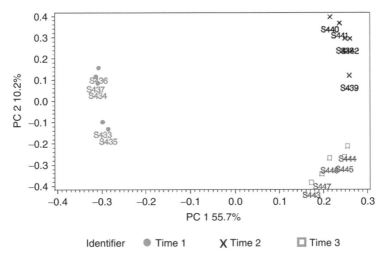

Figure 5.1   Influence of protocol changes and shift over time on gene expression data. Principal
component analysis (PCA) of gene expression data from five replicate UHHR specimens processed
in different batches. The same RNA samples were processed by operator 1 (time 1, gray dots) and
operator 2, using the same protocol but different heating blocks. Operator 2 processed the same
RNA at two different time points, separated by 1 month (time 2, crosses, and time 3, gray empty
squares). Both choice of protocol and processing time point influence gene expression data, though
not to the same degree.

whole-genome custom array (more than 50 000 probes) from time 1 differ from those at time 2 and time 3 in several important ways. First, two operators were involved: one who processed at time 1, another one who processed at times 2 and 3. While carrying out root-cause analysis of the differences, it was discovered that the two operators used heat blocks from two different manufacturers during their processing! This was the only significant deviation found in the protocol. Operator 2 processed at times 2 and 3, separated by one month. However, even running the same protocol with the same operator, though separated by one month, we see that time 2 and time 3 samples are still separable in the second principal component. Intensities within the time 2 and time 3 batches generally have very low coefficients of variation (median CV $\leq 5\%$, robust multi-array average (RMA) as summarization/normalization method). This allows for the detection of small biases. However, the biological effects in many experiments are usually larger than either variability due to repetition within time or between times.

5.2.2 Same Protocol, Different Sites (Study 1)

We designed an experiment in which we focus on the technical sources of variability following the RNA extraction. As depicted in Figure 5.2(a), five male rats (*Rattus norvegicus*) were treated with a PPARα antagonist orally once per day for two weeks, and five matched vehicle-treated male rats of the same age served as controls. After sacrifice, liver samples were snap-frozen and put into RNA-later (Ambion, Inc.) until further use. Animal housing and maintenance, treatment, liver dissection and RNA extraction were all performed in one facility, each step by a single, highly qualified technician. Total RNA was obtained by standard procedures as described elsewhere (Chomczynski and Sacchi 1987). An aliquot of 60 µg total RNA was sent to and processed by 12 technicians (operators). Explicitly, each operator followed the same experimental standard operating procedures. Labeling was done with a starting amount of 5 µg, using the Affymetrix labeling kit, following the manufacturer's instructions. RNA was hybridized to Affymetrix RAE230 arrays, and the resulting CEL files RMA-processed (with background correction, probe summarization using median-polish and quantile normalization). The data are available on the book companion website www.the-batch-effect-book.org/supplement. Figure 5.2(b) shows the boxplots (25–75% quantiles) of the signal intensities of the individual arrays after RMA normalization. As expected, the boxplots of all arrays/batches are similar. Lab5_AF has a slightly larger spread of signal intensities than the other batches, both for the untreated group and for the treated group. In Figure 5.2(c) it becomes more apparent that Lab5_AF (thick black line) is different from the other batches, as the distribution of the mean signal intensity values deviates from the intensity distribution of the other batches below $\log_2$ intensities of 8. As can be seen in the principal components analysis (PCA) score plot (Figure 5.2(d)), the main source of variation (along the first principal component) is the treatment factor, but there is an appreciable variance in the data coming from the different sites and the different technicians, along the second principal component. Multidimensional scaling (MDS; Grimm and Yarnold 2001) is another helpful statistical method for visualizing high-dimensional data. The nonparametric version of MDS used focuses on the representation of the (smaller) differences between individual batches. This explains the difference with respect to the PCA score plot and provides additional insight. In the MDS plot in Figure 5.1(e) the individual batches are more clearly separated compared to

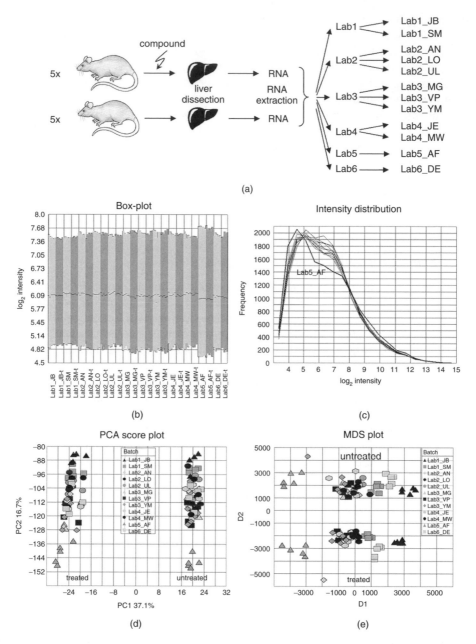

Figure 5.2 Inter-laboratory variation of gene expression caused by different handling of RNA- and microarray processing procedures. (a) Experimental design. (b) Box plots (25–75% quantiles) of signal intensities of individual arrays. (c) Intensity distributions of group means (by operator) of the untreated samples. (For simplicity the similar results for the treated samples are not shown). Lab5_AF is highlighted as the thick black line. (d) PCA score plot of 120 Affymetrix RAE230 microarrays from 12 operators in six laboratories. Along PC1 the treatment is the biggest source of variability. (e) MDS plot of the same samples as in (d). Along D1 the laboratories are the biggest source of variability.

the PCA score plot. This allows for a better qualitative estimation of the relative sizes of the overall batch effects.

As indicated by the hierarchical cluster analysis of the mean expression values (averaged over all samples per batch) of untreated samples shown in Figure 5.3(a), there is a trend that intra-laboratory variance (i.e. inter-operator variance in a single laboratory) is smaller than inter-laboratory variance (the result for the 'treated' group is very similar, and therefore not shown here). This makes sense as performance of an identical protocol within a laboratory, although handled by different technicians, should be more similar than the performance of different protocols.

What is the effect of the handling by different operators on gene expression levels? We initially address the issue of comparability of lists of differentially expressed genes per batch, the concordance between each batch and Lab1_JB as common reference. We compared lists of 1, 5, 10, 50, 100, 250, and 500 probe sets with largest fold changes between the means of treated and untreated animal groups, after application of a p-value filter ($p < 0.01$) to probe sets with a minimum mean expression of 6 (in $\log_2$ units) in the control group. The concordance of each batch with Lab1_JB as reference is between 67% and 80% for lists containing more than 10 probe sets, which shows that there is good agreement between the lists of differentially expressed genes (data not shown).

Next we analyzed how many probe sets within a range of signal intensities have a higher or lower mean signal intensity in one batch compared to another batch. The minimum mean ratio (i.e. fold change) should be larger than a threshold value which is usually thought of as a good starting point for gene discovery analyses, so we chose 1.2 or 1.5 (in either direction). We compared two very similar batches, Lab3_VP and Lab3_YM, treated and untreated samples separately (Figure 5.3(b)). We applied a fold change threshold of 1.2 which is fairly low. When using 1.5 as the lower cutoff there are hardly any probe sets changed (not shown). For a fold-change cutoff of 1.2, only a small percentage of probe sets per signal intensity range are affected at all (Figure 5.3(c), (d)). Next we turned to dissimilar batches and compared Lab5_AF and Lab1_JB, treated and untreated samples separately (Figure 5.3(e)). In each case the distribution of affected probe sets across signal intensity ranges is not uniform. The results for the dissimilar batches shows that only a few probe sets with low or very high signal intensities are affected, while up to 27% of all probe sets in the intensity range 8–9 ($\log_2$ scale) have a more than 1.5-fold difference between batches (Figure 5.3(f), (g)). In absolute numbers, looking at the untreated samples only (numbers for treated are very similar), there were 282 probe sets (i.e. 1.8% of all probe sets on the RAE230 array) affected by processing of the RNA in very similar batches, and 2940 probe sets (18.5% of all probe sets) had a fold change of at least 1.5 between two very dissimilar batches. Of those 2940 probe sets, 1168 have a p-value smaller than 0.05, and a signal intensity larger than 7. This means that 1168 probe sets would pass all standard criteria of being 'differentially expressed' between two data sets. Remember that the two data sets are based on aliquots from the same RNA!

5.2.3 Same Protocol, Different Sites (Study 2)

We wanted to see whether our findings can be also be observed in another data set with a similar setting. In 2006, an international initiative driven by the US Food

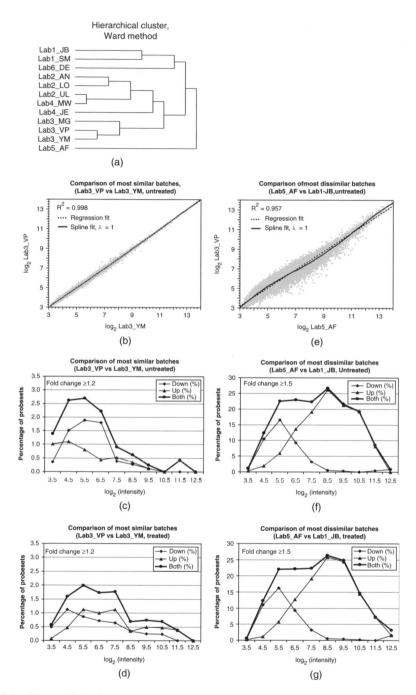

Figure 5.3 Hierarchical cluster analysis (Ward's method) of mean expression data of untreated rat liver samples (a). The percentage of probe sets with relevant fold changes between very similar batches ((b)–(d)) is smaller than between very dissimilar batches ((e)–(g)). The number of affected probe sets within a signal intensity range is given relative to the total number of probe sets within the signal intensity range.

and Drug Administration, the MicroArray Quality Control Consortium (MAQC), published data in an attempt to address the question whether microarray data sets (based on an identical set of samples) generated on different platforms and in different laboratories would yield reproducible results (Shi *et al.* 2006; http://www.fda.gov/science/centers/toxicoinformatics/maqc)). One of the MAQC data sets consisted of replicates of aliquots of two human reference RNAs and mixtures thereof: 'Universal Human Reference RNA, UHRR' (sample A, provided by Stratagene)) and 'Human Brain Reference RNA, HBRR' (sample B, provided by Ambion). The samples were processed at six sites on the Affymetrix platform, each site generating one HG-U133plus2 array per sample from five replicates (data are available from http://www.ncbi.nlm.nih.gov/geo; accession number GSE5350). Here we limit ourselves to the presentation of the UHRR data, as the results for the HBRR data are very similar. In a PCA score plot using all probe sets we can clearly see that site 6 is more different from sites 1−5 than those are from each other (Figure 5.4(a)). Unfortunately, the source of this outlying behavior of site 6 could not be identified (L. Shi, personal communication). We also noticed that the data generated at site 4 have a larger variance than the other sites, possibly due to technical issues, making it an outlier is this aspect. Probe sets which are affected most are in the intensity range of about 6−9 (on the $\log_2$ scale), which is a region whose expression data tend to be trusted as they are above array background noise and below saturation (Figure 5.4(b)).

When we plot between-batch fold changes versus signal intensity we obtain a picture similar to that obtained in the previous analysis. There are only very few affected probe sets when very similar batches are compared (Figure 5.4(c)), but a large number of probe sets in the comparison of dissimilar batches (Figure 5.4(d)). There appears to be a difference in the profiles of probe sets with higher/lower intensity in the reference batch in the UHRR data set compared to the rat data set, since in the UHRR the lines barely cross (Figure 5.4(d)) while they intersect in the rat data set (see Figure 5.4(a), (b)). This observation indicates that different batches may have different profiles.

5.2.4 *Batch Effect Characteristics at the Probe Level*

We wanted to investigate how a batch effect is characterized on the probe level, in contrast to the probe set level investigated so far. In an inter-laboratory comparison of gene expression data, run by Expression Analysis Inc., two laboratories processed identical samples. In this particular case, targets from the same mouse RNA source were prepared and labeled for the Affymetrix U74Av2 mouse chip. A batch effect at the probe level was detected that has been reproduced repeatedly in other contexts: A subset of probes (dark points in the scatterplot in Figure 5.5) was identified that consistently yielded higher intensities in one laboratory than another. All of these probes had a common attribute: each perfect match (PM) probe contains a set of four successive G nucleotides. Since our first discovery of this behavior in 2003, we have occasionally seen this bias occur in other Affymetrix arrays besides this illustration for the U74Av2 mouse chip. It has also been identified independently by others (Upton *et al.* 2008; Zhang, MD Anderson Cancer Center, private communication). Upton *et al.* believe that the tendency of GGGG-containing probes to be brighter than other probes in a probe set is essentially due to

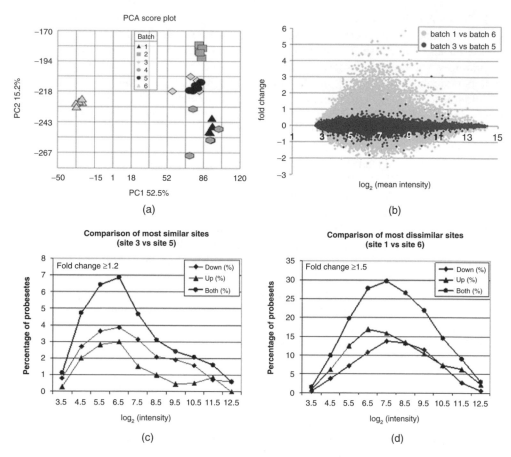

Figure 5.4 Inter-site variation (RNA processing and array hybridization) contributes to the signal intensity changes of identical RNA. (a) PCA score plot of Affymetrix HG-U133plus2 arrays hybridized with five UHRR samples in six laboratories (sites). Site 6 generated data which are markedly different from the other sites. (b) Scatterplot of mean signal intensities of site 1 and 6, and of site 3 and 5. (c), (d) Percentage of probe sets with relevant fold change from one site to another. Note the different y-axis scales.

probe–probe interaction on densely packed arrays. Thus the GGGG probes may form quadruplexes which show abnormal target binding qualities, as the effective association rate of probe and target is proportional to probe density. Many factors influence the stability of the quadruplexes, for instance the presence of monovalent cations or ethanol, which induces the formation of quadruplexes, storage conditions of arrays (low/high temperatures), or whether or not an array was heated prior to hybridization. In the example presented (Figure 5.5), the protocols may have been slightly different, but possibly other factors such as enzyme lot or array storage condition may have been slightly different as well. As the data were generated in 2003, a root-cause analysis is not possible anymore.

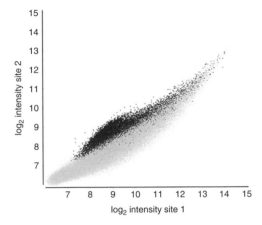

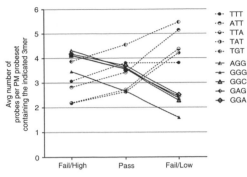

Figure 5.5 Approximately 4% (8500 of 197 000) of the perfect match probes on the Affymetrix U74Av2 array contain four consecutive G nucleotides. These probes are highlighted in black. The 'multiple-G' effect was detected in data sets hybridized in both lab 1 and lab 2. These results suggest that the hybridization changes are related to a protocol difference during target preparation.

Figure 5.6 Nucleotide content of probes can be a factor in batch effects. In this example of an experiment with Affymetrix HG-U133A arrays, probe sets with higher signal intensities in lab 1 compared to lab 2 were enriched in PM probes with higher content of T nucleotides, while probe sets with lower signal intensities in lab 1 were enriched in PM probes with higher G nucleotides.

Depending on the brightness of other probes which make up the probe set, the GGGG-containing probes may receive an outlier call (other probes have low brightness) or may contribute stronger to the calculated signal intensity than other probes (gene is highly expressed and other probes have strong brightness). Thus, the probe sequence turns out to be a factor which contributes to the calculated signal intensity, which is dependent on protocol, array storage condition, and possibly other factors, which are not yet identified.

In yet another experiment, six identical replicates of RNA from the same source (UHRR, Stratagene Inc.) were prepared and hybridized to Affymetrix HG-U133A GeneChips at two different laboratories (Figure 5.6). One laboratory was used as a reference lab (lab 1) and the other laboratory was examined for consistency of absolute expression intensities (lab 2). Average signal intensity differences between the two laboratories were assessed for statistical significance using a t-test with adjustment for multiple testing. Each Affymetrix probe set was graded as 'pass' if the t-test did not indicate a significant difference. Fold change was not used as a separate criterion, although it plays an important part in the t-test itself. Due to the conservative p-value adjustment ($p < 0.000001$), very small fold changes (e.g. FC < 1.1) would rarely be flagged as significant. Roughly 5000 of the more than 22 000 HG-U133A probe sets did not pass the t-test. When root cause analysis was performed to understand this effect, a pattern was detected in those probe sets whose absolute intensity was consistently lower or higher in lab 2 than in lab 1. In particular, probe sets with probes enriched for the T nucleotide were found to have consistently lower signals in lab 2 than in lab 1. Likewise, probe sets with probes enriched for the G nucleotide were found to have consistently higher signal in lab 2 than in lab 1. Figure 5.6 illustrates how nucleotide content/sequence can be an important factor in batch effects. The figure shows that, depending on the frequency of a particular sequence of three consecutive nucleotides within the PM probes included in a probe set, labs tend to show different signal values for that probe set. For example, when one examines those probe sets where lab 2 had greater (higher) signal than lab 1, the probes within the probe set had an average of 4 PM probes containing (for example) AGG. However, when the signals were equivalent, the probe sets had roughly 3.5 PM probes containing AGG and less than 2.5 PM probes with AGG when lab 2 probe sets had lower signal than lab 1. There were similar results for other 3-mers where there were at least two Gs. However, when we look at 3-mers containing multiple Ts, the results are reversed. Hence, again, probe sequence appears to play a big role in the observed inter-laboratory batch effects of many probe sets.

5.3 Conclusion

By using real microarray data from a toxicogenomics experiment and several experiments where RNA was reprocessed multiple times under identical or slightly different conditions, we see that batch effects can be substantial and greatly influence the data of microarray experiments and their interpretation.

The size and the direction of batch effects cannot be predicted. Our results show that batch effects are not unidirectional (i.e. all affected probe sets are biased in the same direction) or symmetrical (same number of probe sets biased positively and negatively). Additionally, the number of affected probe sets and the size of the batch effects can vary greatly. Another important finding is the observation that the proportion of affected probe sets depends on their expression intensity. The fact that we have found different signatures of batch effects in the experiments we have investigated points to a phenomenon which may not be extrapolated to other experiments and causes of bias. We suggest that each experiment should be investigated separately in this respect. Other findings indicate that the probes most affected by batch effects have certain characteristic nucleotide sequences.

6

Bioinformatic Strategies for cDNA-Microarray Data Processing

Jessica Fahlén, Mattias Landfors, Eva Freyhult, Max Bylesjö, Johan Trygg, Torgeir R Hvidsten, and Patrik Rydén

Abstract

Pre-processing plays a vital role in cDNA-microarray data analysis. Without proper pre-processing it is likely that the biological conclusions will be misleading. However, there are many alternatives and in order to choose a proper pre-processing procedure it is necessary to understand the effect of different methods. This chapter discusses several pre-processing steps, including image analysis, background correction, normalization, and filtering. Spike-in data are used to illustrate how different procedures affect the analytical ability to detect differentially expressed genes and estimate their regulation. The result shows that pre-processing has a major impact on both the experiment's sensitivity and its bias. However, general recommendations are hard to give, since pre-processing consists of several actions that are highly dependent on each other. Furthermore, it is likely that pre-processing have a major impact on downstream analysis, such as clustering and classification, and pre-processing methods should be developed and evaluated with this in mind.

6.1 Introduction

Pre-processing of cDNA-microarray data commonly involves image analysis, normalization and filtering. Over the last decade, a large number of pre-processing methods have been suggested which makes the overall number of possible analyses huge (Mehta *et al.* 2004). Pre-processed data are always used in some type of downstream analysis. Such analysis ranges from identification of differentially expressed genes (Lopes *et al.* 2008; Stolovitzky 2003), through clustering, classification and regression analysis

Batch Effects and Noise in Microarray Experiments: Sources and Solutions edited by A. Scherer
© 2009 John Wiley & Sons, Ltd

(Alizadeh *et al.* 2000; Roepman *et al.* 2005; Ye *et al.* 2003; Zervakis *et al.* 2009), all the way to systems biology and network inference (Lorenz *et al.* 2009).

The choice of pre-processing method affects the downstream analyses (Rydén *et al.* 2006; Ye *et al.* 2003). Hence, pre-processing is important and should be selected with care. The ultimate goal of pre-processing is to present the data in a form that allows modeling of biologically important properties. In this chapter, we discuss how pre-processing affects the result of various analyses. Our aim is not to present an overview of pre-processing methods or to compare methods, but to show the principal effect of applying some commonly used approaches.

As for all experimental procedures, the microarray technology measures not only the desired biological variation but also the technical variation introduced by the experiment. For example, the technical variation can be caused by cell extraction, labeling, hybridization, scanning and image analysis. The technical variation might be systematic and introduce bias, or behave as pure noise. Pre-processing aims to remove this undesired variation.

Although the number of sources contributing to the technical variation is large, it is still possible to describe the merits of different analyses. In order to do this we will use spike-in data to estimate some measures of interest (sensitivity and bias) and use various plots (the intensity–concentration (IC) curve and *MA* plot) to describe the systematic variation. In this introductory section we introduce spike-in data, the IC curve, the *MA* plot and some key measures. In Section 6.2 we show how the sensitivity and bias are affected by various pre-processing methods. In Section 6.3 we discuss how pre-processing methods may influence downstream analyses and present a tumor data example that illustrates how different pre-processing methods influence a cluster analysis. A discussion and the major conclusions are presented in Section 6.4.

6.1.1   Spike-in Experiments

In a spike-in experiment, all the genes' RNA abundances are known. The advantage of using spike-in data for investigating the effect of pre-processing methods is, in contrast to ordinary experiments, that all key measures can be estimated. A commonly used alternative is to simulate realistic microarray data, but this is a very difficult task. The simulation has to build on various model assumptions that generally cannot be validated. Furthermore, spike-in data have the advantage that they go through the same experimental steps as an ordinary experiment and are therefore subject to the same technical variation.

We consider data from eight in-house produced spike-in cDNA-microarrays (the Lucidea experiment). The arrays were in-house produced cDNA-arrays consisting of 20 clones from the Lucidea™ Universal ScoreCard. Each clone was printed 480 times in 48 identically designed sub-grids. Eight Lucidea arrays were hybridized with labeled preparations of Lucidea Universal ScoreCard reference and test spike mix RNA, along with total RNA from murine cell line J774.1 (data available on book companion site www.the-batch-effect-book.org/supplement). The arrays had approximately 6000 nondifferentially expressed (NDE) genes and 4000 differentially expressed (DE) genes. The NDE genes had RNA abundances ranging from low to very high. The DE genes were either threefold or tenfold up- or down-regulated, with either low or high RNA abundances. For further details, see Rydén *et al.* (2006).

6.1.2   Key Measures – Sensitivity and Bias

We consider cDNA-microarray experiments where two populations are compared and where the aim is to identify and describe biological differences between the populations. An experiment is characterized by its ability to identify DE genes and correctly estimate the regulation of the DE genes (i.e. sensitivity and bias).

The difference in a gene's expression between the two populations is estimated by the average log-ratio (taken over all arrays) and a test statistic is constructed in order to determine how likely it is that the gene is differentially expressed. Genes with p-values below a user-determined cutoff value are classified as DE and the remaining genes are classified as NDE. The cutoff value is determined so that the false discovery rate (FDR) is kept at the user's desired level. The FDR is the proportion of false positive genes among the selected genes. A reasonable FDR is often set at around 5–10%, but this depends on the investigator and the aim of the study. Determining the cutoff value is trivial for spike-in experiments because the gene regulations are known in advance, but obviously much more difficult for ordinary experiments. For spike-in data the experiment's sensitivity (probability of observing a true positive) and specificity (probability of observing a true negative) can easily be estimated for any cutoff value.

Since only a small fraction of the genes are assumed to be differentially expressed the cutoff value will mainly be governed by the NDE genes with the most extreme test statistics (i.e. lowest p-values). We will consider the sensitivity when the specificity is fixed at 99.95%. This corresponds to a FDR around 5–20% when the sensitivity is in the range 20–90% and only 1% of the genes are truly differentially expressed.

How well an experiment is able to predict the true regulation of the DE genes is another important measure for judging the quality of the experiment. The bias of a DE gene is the expected difference between the observed and true regulation. Since it is common practice to transform the intensities using the logarithmic transformation with base 2 we also consider the bias on log scale. The bias for one DE gene is estimated as the difference between the average observed log-ratio (taken over all arrays) and the true log-ratio. To estimate the combined bias for two DE genes is a more delicate task. If we only take the average of the two biases it would be rather misleading. Consider a situation where we have one up-regulated and one down-regulated gene, and that the experiment underestimates all types of regulation. In this case the bias will be negative for the up-regulated gene and positive for the down-regulated gene, but the average might be close to zero. Our solution to this problem is to consider the reflected bias, where all down-regulated genes have their observed and true regulation multiplied by -1. Once this is done we can estimate the combined bias with the average of the reflected biases. This approach allows us to estimate the overall bias, while retaining its direction (i.e. over- or underestimation).

6.1.3   The IC Curve and MA Plot

In a spike-in experiment, all the RNA abundances are known and all genes are designed to have similar properties. It is therefore possible to study the relationship between the logarithm of the genes' RNA abundance (the concentration) and the expected value of the corresponding log-intensities. The expected values are estimated with the average

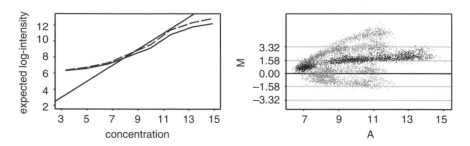

Figure 6.1   IC curves and *MA* plot for the raw data obtained at the 80 scan. The left plot shows the IC curves of the treatment (dashed) and reference (solid) channels. The straight line is the ideal IC curve. The right plot shows the corresponding *MA* plot, were the black dots correspond to NDE genes and the gray dots to DE genes. The horizontal lines represent the true regulation of the genes. Clearly, the data are affected by the background since we have no intensities below 6.

log-intensities taken over a set of arrays and a set of replicates (genes with the same concentration). The intensity–concentration (IC) curve illustrates this relation (e.g. Figure 6.1) and is a powerful tool to study the effects of applying different pre-processing methods. In an ideal situation, the IC curve is a straight line through origin and with slope equal to one; that is, doubling the concentration results in doubled log-intensities. However, the estimated IC curves commonly deviate from the ideal curve. Typically, raw data produce IC curves that are S-shaped, and due to systematic variation the observed slopes are often less than one. The change in slope introduces bias in the observed log-ratios of the DE genes, so that the magnitudes of the regulations of DE genes are underestimated. Note that the bias increases with the magnitude of the true regulation.

To illustrate the distribution of the extreme NDE genes, those that are highly responsible for the sensitivity, we use the *MA* plot, where the log-ratios (*M*) are plotted against the average log-intensities (*A*); see, for example, Figure 6.1. In the ideal situation the NDE genes should be centered at zero and the DE genes at their true log-ratios. In order to achieve 100% sensitivity it is sufficient that the variation is so small that the NDE genes and DE genes are completely separated.

6.2   Pre-Processing

Pre-processing in a wide sense includes image analysis, selection of data (if we have multiple scans), normalization, and filtration. In particular, normalization aims to reduce the systematic bias while preserving the biological variation. In this section we study how scanning procedures, normalization and filtration affect the overall bias (reflected bias) and sensitivity. Throughout this chapter we use the IC curves and *MA* plots from the Lucidea experiment to illustrate the changes in bias and sensitivity. For clarity, the IC curves are based on data from one array while the *MA* plots are based on the aggregated Lucidea data. In all examples, the *B* statistic (Lönnstedt and Speed 2002) was used as the test statistic and the specificity was kept at 99.95%.

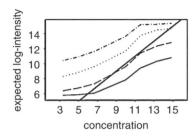

Figure 6.2 IC curves for the Lucidea raw data scanned at different scanner settings; the 70 scan (solid), the 80 scan (dashed), the 90 scan (dotted) and the 100 scan (dashed and dotted).

6.2.1 Scanning Procedures

The location of the IC curves is affected by the scanner intensity. In the Lucidea experiment the arrays were scanned at four settings: 70%, 80%, 90% and 100% of the maximum laser intensity and photomultiplier tube (PMT) voltage. Henceforth, these scans are referred to as the 70, 80, 90 and 100 scans. The IC curves for the four settings are shown in Figure 6.2. Generally, the number of saturated spots will increase with the scanner settings. On the other hand, by lowering the scanner settings we will increase the number of not-found spots (i.e. genes that cannot be separated from the background noise and are flagged as not found during the image analysis). The relation between the scanner settings, the amount of saturated and not-found spots for the Lucidea data is presented in Table 6.1. The location of the IC curves is also affected by the scanner intensity, but the parallel shift of the IC curves is generally irrelevant in constructing unbiased estimators of the log-ratios.

6.2.2 Background Correction

A common problem when measuring optical signals is that the raw intensities are affected by background errors. There are several sources that contribute to background errors, such as cross hybridization, unbound RNA/DNA, dust, stray light and Pmt noise (dark noise). An observed intensity is commonly modeled as the sum of the 'desired' intensity and the background error, where the two variables are independent. Under these assumptions it is

Table 6.1 Percentages of not-found and saturated spots for one array in the Lucidea experiment at four different scanner settings.

	70 scan	80 scan	90 scan	100 scan
Saturated spots (%)	0	0.1	9	19
Not-found spots (%)	52	49	44	41

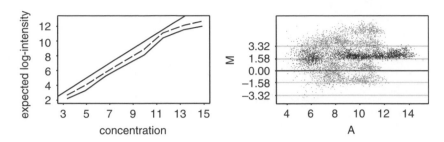

Figure 6.3 IC curves and *MA* plot for background corrected 80 scan Lucidea data. The left plot shows the IC curves of the treatment (dashed line) and reference (solid line) channels. The straight line is the ideal IC curve. The right plot shows the corresponding *MA* plot, were the black dots correspond to NDE genes and the gray dots to DE genes. The horizontal lines represent the true regulation of the genes.

clear that the background errors mainly affect weakly expressed genes. This can typically be seen in the IC curves; see, for example, Figure 6.1. Moderately and highly expressed genes are only slightly affected, but importantly the background causes a reduction in the slope of the IC curve.

Background correction methods aim to remove the background from the raw intensities. In our examples we have applied local background correction where the spot's local background (measured around the spot) is subtracted from its intensity (Eisen 1999). In Figure 6.3 we see how the background correction straightens out the IC curves. Thus, the correction reduces the overall bias, but interestingly it will entail a prominent increase in the variance of the log-ratios. In particular, background correction will increase the number of extreme log-ratios from the NDE genes. The *MA* plots for the non-background-corrected and background-corrected data clearly show this drawback (Figures 6.1 and 6.3). The increased variance makes it harder to detect DE genes (lower sensitivity); see, for example, Qin and Kerr (2004) and Rydén *et al.* (2006). In Table 6.2 the bias and the sensitivity for data with and without background correction are presented. The table highlights the trade-off between sensitivity and bias: the background correction reduced the sensitivity from 68% to 41%, but it also reduced the bias from −0.8 to −0.2. A reflected bias equal to −0.8 (−0.2) tells us that 57% (87%) of the magnitude of the true regulation of the DE genes is observed. An additional problem is that background correction produce a large number of negative intensities; this will be discussed in Section 6.2.5.

The increased variance is often explained by the fact that the background always is estimated with some error, and that we introduce additional variance in the subtraction step. Evidently there is some truth in such a statement, but in fact the variance will increase even though we remove the true background. The fact that background correction commonly result in increased variance and decreased bias can be explained by the following theoretical argument. Assume that X and Y are two positive and independent random variables such that

$$E\left[\log\left(\frac{X}{Y}\right)\right] = \mu > 0, \quad \text{Var}\left[\log\left(\frac{X}{Y}\right)\right] = \sigma^2.$$

Table 6.2 Sensitivity at 99.95% specificity and reflected bias for different normalization methods. Data from the Lucidea 80 scan were used and two types of background correction were considered: no correction (No) and local background correction (Local). Three types of dye normalizations were considered; no dye -normalization (No), *MA* normalization (Global) and print-tip MA normalization (Spatial). The *B*-test was used for all normalizations.

Background correction	Dye normalization	Sensitivity (%)	Reflected bias
No	No	18	*
Local	No	17	*
No	Global	45	−0.8
Local	Global	31	−0.2
No	Spatial	68	−0.8
Local	Spatial	41	−0.2

*Reflected bias is designed for data centered at zero and is not a sensible measure for data that have not been dye normalized.

Then, for any positive constant a, we have

$$E\left[\log\left(\frac{X+a}{Y+a}\right)\right] < \mu, \quad \mathrm{Var}\left[\log\left(\frac{X+a}{Y+a}\right)\right] < \sigma^2.$$

Adding a positive constant corresponds to adding a positive background to the 'desired' intensities. An intuitive explanation for the increased variance is that the background-corrected intensities from weakly expressed genes behave as random noise with mean close to zero. When constructing the log-ratios we get division by values close to zero and as a consequence some extremely high ratios.

Several techniques to improve the removal of the background errors have been suggested (Efron *et al.* 2001; Kooperberg *et al.* 2002; Yang *et al.* 2002a; Yin *et al.* 2005). For a more detailed description and comparison of different background correction methods, see Ritchie *et al.* (2007).

6.2.3 Saturation

The current scanners have a limited resolution (limited to 16-bit images), which causes highly expressed genes to have saturated intensities (i.e. intensities that are affected by pixel values that are truncated at the maximum value $2^{16} - 1$). This causes a censoring of the highly expressed genes, which appear as the upper knee in the IC curves (Figure 6.1). This decrease in slope affects the bias of the highly expressed DE genes. Contrary to the background, which affects all genes, the saturation only affects genes that are expressed at high levels. Thus, correcting for saturation reduces the bias of highly expressed DE genes. How this correction affects sensitivity is less clear, but if a large proportion of the DE genes have saturated intensities, then it is likely that the correction will increase the overall sensitivity. The bias caused by saturation can be avoided by considering data from a low scanner setting (Table 6.1). However, the background problems are generally high at low settings (Figure 6.2). A solution is to combine data from several scanner settings (Bengtsson *et al.* 2004; Dudley *et al.* 2002; Lyng *et al.*

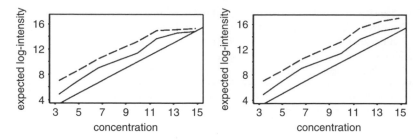

Figure 6.4 IC curves for background corrected 100 scan Lucidea data before (left) and after (right) correction of saturated intensities. The plot shows the IC curves of the treatment (dashed) and reference (solid) channels. The straight line is the ideal IC curve. Linear scaling, combining data from the 80, 90, and 100 scans, was used to remove the systematic variation caused by saturation.

2004). Figure 6.4 shows the IC curves before and after saturation correction using linear scaling of data from three scanner settings similar to what was described in Dudley *et al.* (2002).

6.2.4   Normalization

6.2.4.1   Dye Bias: General Considerations

In a cDNA-experiment there are experimental differences between the populations; for example, cells are extracted separately, the samples are labeled with different dyes, and different wavelengths are used during scanning. These differences influence the background and saturation biases, but also introduce an array and dye-specific bias. The array-specific bias is generally characterized by a global shift in the intensity levels of each microarray element. The dye-specific bias can be thought of as the difference between the populations' IC curves after all the background and saturation bias have been removed. Dye normalization aims to normalize the populations' intensities into 'a common scale', such that the populations' IC curves coincide. The normalized IC curve can be regarded as the 'average' of the original IC curves (Figure 6.5). We stress that dye normalization does not remove background and saturation bias; it just puts the data on a common scale.

6.2.4.2   Spatial Dependency

In order to put the data on a common scale, dye normalization methods generally normalize the data so that the log-ratios of the NDE genes are centered at zero; see, for example, Figure 6.6. Some methods assume that the dye differences are homogeneous (Dudoit *et al.* 2002; Bolstad *et al.* 2003), and other methods assume that there is a spatial dependency over the arrays (Wilson *et al.* 2003; Yang *et al.* 2002c). Such spatial effects may be caused by uneven hybridization and washing. Thus, using methods that account for spatial dependency will improve the normalization and increase the overall sensitivity. The improvement can be significant; see, for example, Table 6.2 where the global

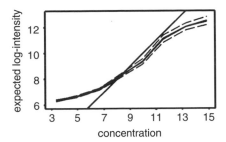

Figure 6.5 IC curves for the 80 scan Lucidea data before (dashed) and after (solid) dye normalization. Note that, here both the channels are described by the same type of lines and that IC curves of the channels' normalized data are very close to each other. The data were normalized using the print-tip *MA* normalization.

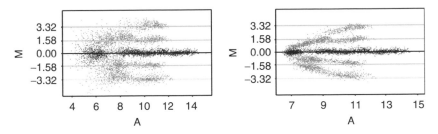

Figure 6.6 The *MA* plots for dye-normalization (a) without background correction and (b) with background correction for the 80 scan Lucidea data. The black dots correspond to NDE genes and the gray dots to DE genes. The horizontal lines represent the true regulation of the genes.

MA normalization (Dudoit *et al.* 2002) did have considerably lower sensitivity than the print-tip *MA* normalization (Yang *et al.* 2002c).

6.2.4.3 OPLS Normalization for Modeling of Array and Dye Bias

An alternative to traditional within-array normalization methods would be to include information across multiple arrays in an experiment. This can be helpful for identifying general properties of the array and dye biases. One approach towards multi-array normalization uses the orthogonal projections to latent structures (OPLS) regression method (Trygg and Wold 2002; Bylesjö *et al.* 2007). In OPLS normalization, the design matrix of the experiment (describing the biological background of the samples) is employed to identify systematic variation independent of the design matrix. This is intuitively appealing since it ensures that no covariation in the experiment related to the design matrix will be removed. To do this, OPLS normalization requires a balanced design in order to separate the different sources of variation. For the Lucidea experiment, all treated samples are labeled using one dye and all reference samples using another dye; hence the dye effect and the treatment effect are confounded in the design matrix. In such a design, OPLS normalization is not generally applicable since removing the dye effect (unwanted batch effect) would also imply removing the treatment effect (endpoint of interest).

6.2.5 Filtering

In any experiment a large proportion of the genes will not be expressed or will be expressed at very low concentrations. Their intensities will be on the level of the background noise and most of them are not found in the image analysis. If background correction is applied, several of the weakly expressed genes that are found will have negative intensities after the correction. Henceforth, spots that are either not found or have at least one negative intensity are referred to as flagged spots. Here we present three filtration methods that handle flagged spots: complete filtering (treating the flagged spots as missing values), partial filtering (giving all flagged spots a small user-defined value, so that their log-ratios are set to zero), and censoring (which is a generalization of partial filtering). In censoring all flagged spots, as well as spots with very low intensities, are given a small user-defined value.

A drawback with complete filtering is the loss in efficiency in the downstream analyses. In particular, if the number of arrays is small, and if background correction is applied, then several of the weakly expressed genes will only have a small number of observed log-ratios. Just by chance some of these genes may get very low p-values, resulting in a low sensitivity. A common solution is to remove genes with less than k observed log-ratios. For some k-values this will increase the sensitivity. Unfortunately, this leaves us with the difficult problem of choosing the number k.

Partial filtering is based on the assumption that the majority of the flagged spots are due to the fact that the genes are not expressed in any of the populations and that their true log-ratios are zero. In comparison to complete filtering it reduces the influence of weakly expressed NDE genes and suppresses the log-ratios of the DE genes. For background-corrected data this results in a higher sensitivity and bias compared to complete filtration (Table 6.3).

Table 6.3 Sensitivity at 99.95% specificity and reflected bias for different filtering methods. Data from the Lucidea 80 scan were used and two types of background correction were considered: no correction (No) and local background correction (Local). Three types of filtering methods were considered: complete, partial and censoring (with a minimum value equal to 64). The B-test and print-tip MA normalization were used for all normalizations.

Background correction	Filtering method	Sensitivity (%)	Reflected bias
No	Complete	68	−0.8
Local	Complete	41	−0.2
No	Partial	65	−1.0
Local	Partial	74	−0.6
No	Censoring	68	−0.8
Local	Censoring	78	−0.5

Censoring intensities of the flagged spots, as well as the low intensities (i.e. intensities lower than some value c), are set to a user defined minimum value c. Censoring can be very powerful, but it is an open problem how to determine the minimum value c. For background-corrected data censoring can be regarded as a type of reversed background correction and consequently might result in both increased sensitivity and larger bias (Table 6.3).

6.3 Downstream Analysis

Pre-processing of microarray data is, as the name suggests, a prerequisite for further downstream analysis. Identification of DE genes (often referred to as gene selection or feature selection) is usually an integral step in all downstream analyses. Due to the large number of genes compared to the number of observations, gene selection is essential in order to avoid overfitting in subsequent model induction methods (Hawkins 2004). In this section we discuss methods for gene selection and provide an example of how pre-processing of a real-world microarray data set affects a downstream analysis such as hierarchical clustering.

6.3.1 Gene Selection

Commonly, downstream analysis aims to identify genes or groups of genes that are affected by the treatment. The first step is to rank the genes by using some test procedure. Genes with p-values below some cutoff value are classified as differentially expressed. Here, the cutoff value is commonly determined so that the FDR is controlled at a reasonable level (Benjamini and Hochberg 1995). The list of classified DE genes is generally filtered further using gene ontology (Ashburner *et al.* 2000) or other sources of biological knowledge. The genes are then verified to be differentially expressed by other methods, such as quantitative real-time polymerase chain reaction.

Although the sensitivity is highly dependent on the choice of test, no explicit relation can be given in general. This is because the relative merits of the methods are much dependent on the design of the experiment (including the number of arrays) and the pre-processing. Because of these dependencies the relative merits have at the time of writing not been exhaustively studied. It has, however, been shown that the classical t-test performs relatively poorly for microarray data likely due to the small number of observations (arrays) (Qin and Kerr 2004). Several more complex approaches have been adapted to microarrays to improve the tests. These approaches include stabilization of the sample variance (shrinkage), estimation of the distribution under the null hypothesis through resampling, and Bayesian approaches (Baldi and Long 2001; Lönnstedt and Speed 2002; Tusher *et al.* 2001).

6.3.2 Cluster Analysis

Ye *et al.* (2003) presented a study of hepatitis B virus-positive metastatic hepatocellular carcinomas. The study includes 87 tumor samples, with 65 samples from patients with

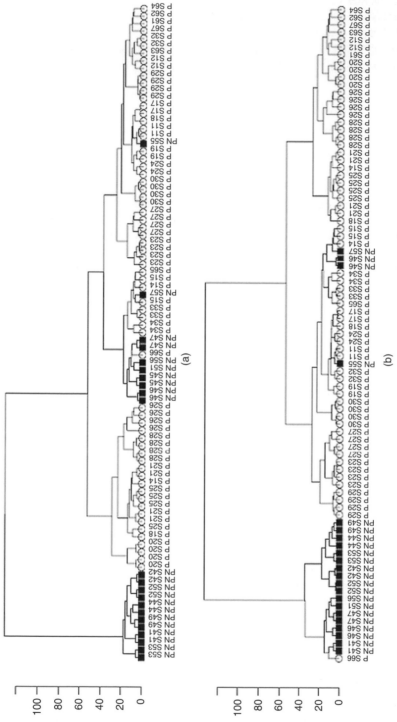

Figure 6.7 Clustering results of the Ye data. The dendrograms show the results of applying Ward's hierarchical clustering after print-tip *MA* normalization (a) without background correction and (b) with background correction. The leaves in the dendrogram are marked with P or PN, depending on the class they belong to, and a number unique to each patient.

metastasis (samples taken from primary and metastatic tumors), class P, and 22 samples from patients with no metastasis (samples taken from primary tumor), class PN.

As Ye points out in his publication, it is very difficult (or even impossible) to separate P from PN samples unless a gene selection method taking the class information into account is used to identify a set of DE genes.

A descriptive analysis of the raw data suggested that there were systematic differences within and between the arrays in the experiment. In addition, it was evident that the background errors where rather large. Therefore, we compared hierarchical clustering results for two different normalizations: print-tip *MA* normalization (Yang *et al.* 2002c) in combination with background correction and without background correction. After normalization, a gene selection method using the class information (P and PN) was employed (i.e. a modified *t*-test (Baldi and Long 2001) was calculated to test the difference between the two classes) and the 100 most differentially expressed genes were selected. In order to use the same gene set in the clustering for both normalizations (background and no background correction), the intersection of the two gene selections was computed. The intersection consisted of 75 genes and these were used in the following hierarchical cluster procedure using Ward's method. Before the actual clustering the data were standardized so that each gene was transformed to have mean 0 and standard deviation 1.

As can be seen in Figure 6.7, the choice of normalization in this example has an obvious effect on the clustering method's ability to separate the two cancer classes.

Pre-processing is likely to affect the cluster analysis. However, more research is needed in order to draw general conclusions.

6.4   Conclusion

Pre-processing is important since different pre-processing methods can lead to different biological conclusions after downstream analysis. Unfortunately, there are numerous alternatives when it comes to pre-processing and there is no universal best method. The first question that needs to be addressed is: what is the aim of the study? If the main objective is to screen for potentially interesting genes, then sensitivity is the top priority and pre-processing methods should be selected accordingly. That said, it is still important to be aware that choosing a pre-processing method that maximizes the sensitivity generally leads to underestimated gene regulation.

On the other hand, if the plan is to carry out some type of more advanced downstream analysis, such as clustering or classification, then both sensitivity and bias should be considered. As demonstrated, there is often a trade-off between low bias and high sensitivity. For example, methods using local background correction commonly have low bias, but also low sensitivity. On the other hand, the use of partial filtration can give high sensitivity depending on whether background correction has been applied or not, but will also result in a relative high bias. This brings us to our next point: pre-processing consists of many actions that are highly dependent on each other. From a user perspective this is bad news, since they would benefit from simple recommendations like 'do not use background correction'. One of our aims was to demonstrate the complexity of pre-processing in that a method's performance depends on which other methods it is combined with. It is also important to have this complexity in mind when introducing and evaluating new methods. This brings us to our final point: pre-processing is likely to have a major impact on downstream analyses such as clustering, classification, and network inference. However, this is still a largely open question that can only be answered by systematic comparisons of several pre-processing methods, downstream analysis methods and biologically different data sets.

7

Batch Effect Estimation of Microarray Platforms with Analysis of Variance

Nysia I George and James J Chen

Abstract

The vast amount of variation in microarray gene expression data hinders the ability to obtain meaningful and accurate analytical results and makes integrating data from independent studies very difficult. In this chapter, we assess the observed variability among microarray platforms through variance component analysis. We utilize the MicroArray Quality Control project data to examine the variability in a study implemented at several laboratory sites and across five platforms. A two-component analysis of variance mixed model is used to partition the sources of variation in the five microarray platforms. The contribution of each source of systematic variance is estimated for each random effect in the experiment. We demonstrate similar inter-platform variability between many of the platforms. For the platform with the highest variation, we find significant reduction in technical variability when data are normalized using quantile normalization.

7.1 Introduction

Microarray technologies, which allow for the simultaneous measurement of thousands of gene expressions, have become a prominent tool in gene expression analysis. Typically, analysis of microarray data begins with image analysis, followed by normalization, and focuses on the detection of differentially expressed genes under different experimental conditions and other post-array analyses (Wilson *et al.* 2003). The primary goal of a microarray study is to successfully extrapolate information from differential expression and provide greater insight into subsequent biological effects.

Batch Effects and Noise in Microarray Experiments: Sources and Solutions edited by A. Scherer
© 2009 John Wiley & Sons, Ltd

7.1.1   Microarray Gene Expression Data

The purpose of a microarray experiment is generally to compare expression levels between two or more experimental conditions (e.g. normal versus diseased, treated versus untreated). The expression data for these comparisons are collected by measuring mRNA abundance in a one- or two-color experimental design. In two-color (competitively hybridized) arrays, the most common design is to hybridize each experimental sample against a common reference sample on a single microarray (Yang and Speed 2002). The true ratio between intensities of the two samples should be one for each gene when there is no difference in intensity between the two conditions. In one-color microarray experiments, single samples are hybridized to one array. The arrays are designed to measure the absolute levels of gene expression. Thus, two single hybridizations are necessary for comparing two experimental conditions.

7.1.1.1   Sources of Variation

Regardless of the microarray platform, the measured expression signal is a combination of true gene expression intensity and various sources of experimental noise in the data. It is important for analysis to be able to distinguish between noise and true signal in order to accurately assess variation in biological samples. Microarray experiments are multi-step procedures which are vulnerable to numerous sources of variation. Given uniqueness in experimental design and the extensive procedure for collecting and processing data, there exists inherent variability in microarray data analysis.

Generally speaking, any observed variability can be explained by biological, systematic (technical), and residual variation (Novak *et al.* 2002; Churchill 2002). The very nature of microarray technology gives rise to systematic variation. This variability is associated with the assay of mRNA expression in each experimental sample. Sources of systematic variability include, but are not limited to, measurement error associated with the array process, mRNA fluorescence labeling and purification, hybridization conditions, and scanning parameters. Biological variability is independent of the experimental process itself and reflects variation from different RNA sources. In simplest terms, it is defined as natural variation in the population from which the samples are derived. It is specific to the system being studied and may be attributed to genetic background, environmental factors, or stage of development. Lastly, residual variability is sampling or experimental variation that cannot be explicitly accounted for.

7.1.1.2   Types of Replication

Replicated microarray data is essential for generating reliable and accurate expression data and filtering statistical noise. Replication can be carried out at different stages of experiment, which allows the magnitude of microarray variability to be evaluated at a variety of levels. According to Nguyen *et al.* (2002), there are three types of replication in gene expression data: spot-to-spot, array-to-array, and subject-to-subject. Spots are replicated by depositing multiple probes for the same gene on one array. In order to replicate arrays, it is necessary to carry out multiple hybridizations using the same

RNA source. Additionally, one can replicate subjects by sampling multiple individuals. With each type of replication, we are able to measure a different source of variability. We can assess within-array variation with spot-to-spot replication, between-array variation with array-to-array replication, and biological variation with subject-to-subject replication. Within-array and between-array variation are attributed to technical variability. Consequently, by including spot-to-spot and array-to-array replication, we are able to assess measurement error and reproducibility in microarray experiments.

It is impossible to completely eliminate variation in each step of microarray analysis. Instead, researchers resolve to estimate the components of variability in an effort to improve statistical inference (Cui and Churchill 2003b). One of the most practical ways to investigate the variability in microarray data is through analysis of variance components.

7.1.2 Analysis of Variance in Gene Expression Data

Analysis of variance (ANOVA) methods play a significant role in the statistical analysis of single- and multiple-factor experiments. They are particularly useful in microarray data analysis since each intensity expression of an experiment is associated with several factors (e.g. array, experimental treatment, and gene). Often in microarray analysis it is in the researcher's interest to quantify the contribution of each source of variation. This is easily carried out using ANOVA modeling, which allows variance in the data to be separated into meaningful components.

The factors (effects) in an ANOVA model may be fixed or random. Fixed effects are factors whose levels are intentionally set by the experimenter, rather than randomly selected from an infinite population of possible levels. Otherwise, the factor is a random effect. Random factors represent the larger population of values. For example, in microarray data analysis, treatment and time point are typically fixed effects and biological replicate is often considered a random effect. If all the controlled or selected variables for experimental factors have fixed (random) effects then a fixed (random) effects model is appropriate for analysis. A mixed effects model has variables that are both fixed effects and random effects.

Much research has been devoted to fixed effects analysis of microarray data (e.g. Kerr *et al.* 2000a; Kerr and Churchill 2001a; Lee *et al.* 2002). However, any inference from a fixed effects model is restricted to factor levels defined in the experiment. Results may not be generalized to observations or levels outside the specific ones being studied. The fixed effects model assumes only technical variation in microarray data and does not allow for multiple sources of variation (Cui and Churchill 2003a). Alternatively, one may consider a mixed effects model. Generally, mixed effects modeling is implemented to estimate variance components. It is a common way to account for each factor that contributes to the overall variation.

Kerr *et al.* (2000b) demonstrate that ANOVA models can successfully normalize gene expression data and provide estimates of error variability. The concept of extending microarray data analysis to include both fixed and random factors with mixed effects ANOVA modeling began with Wolfinger *et al.* (2001) and Jin *et al.* (2001). Mixed models allow for more precise estimation of sources of variability that are necessary in detecting differentially expressed genes (Churchill 2004). Dobbin and Simon (2002) and Dobbin *et al.* (2003) implemented mixed effects ANOVA models in order to best account for biological replicates. Recently, Juenger *et al.* (2006) used ANOVA to assess sources of variability in *Arabidopsis thaliana* expression data.

7.2   Variance Component Analysis across Microarray Platforms

Issues of replication and reproducibility are of great concern in microarray applications because it is important to determine if one can successfully obtain the same findings from the same biological sample when the experiment is repeated. Ideally, one should find that gene expression levels of a gene from the same sample are more similar to each other than those of the same gene from different samples. An experiment of this magnitude focuses on variability within one microarray platform. However, experimental designs for microarray studies vary in scope. It is often of interest to assess performance using different methods of analysis, across multiple laboratory sites, or across multiple platforms. Each adds new dimensionality to the task of examining potential sources of variability in microarray data.

Variance component analysis is a reliable way to quantify the variation of random factors in a microarray study. It measures the components of the variance of an observation and has an important application in microarray data analysis since technical variability in gene expression data accumulates at each step of the data processing scheme. Typically, the estimated variances are considered in statistical inference. They are also used to determine power and sample size for future studies (Cui and Churchill 2003b).

We employ variance component analysis to study inter-platform comparability. This is important because it is likely that one platform demonstrates consistency in replicate measures, yet does not necessarily produce results that are consistent with other platforms. Furthermore, it has been shown that a microarray platform can significantly affect data analysis (Dobbin *et al.* 2005a; Irizarry *et al.* 2005). Larkin *et al.* (2005) assert that observed differences between microarray platforms are often less explained by biological or technical variation, and more related to the metrics used to evaluate performance. Shi *et al.* (2005a) also observe low concordance among different platforms. However, they attribute this finding to low intra-platform reproducibility and a poor choice of analysis methodology.

Our analyses focus on the variance component analysis of gene expression data from a well-defined, comprehensive study of inter-platform comparability. In this chapter, we use a two-factor ANOVA model to partition the sources of variation in five one-color microarray platforms. We estimate variance components for each platform and evaluate the agreement between technical sources of variation. We also consider two methods of normalization for transforming expression data and evaluate their level of agreement. It is our goal to further assess cross-platform analysis when the data are generated under careful experimental design.

7.3   Methodology

7.3.1   Data Description

We examine variance component analysis of microarray platforms using the MicroArray Quality Control (MAQC) project data (Shi *et al.* 2006). The MAQC project was initiated to address concerns associated with the reproducibility of expression data across multiple sites and the comparability of analyzing identical RNA samples under different microarray platforms. In the project, expression data from four titration samples were measured on seven microarray platforms. Each platform was analyzed at three independent test sites with five technical replicates per target sample.

Ideally, independent analyses from multiple technologies would align in order to facilitate data integration in a single analysis. Several studies substantiate the reproducibility of microarray data obtained using different platforms or processed at different test sites (Shippy *et al.* 2004; Petersen *et al.* 2005). However, other studies (e.g. Tan *et al.* 2003; Marshall 2004) show differentiation in data analysis when RNA samples are processed under different platforms. The unique experimental design provides us the opportunity to assess variance component analysis for data collected under the same experimental setup for different microarray platforms. Data can be obtained from http://www.fda.gov/downloads/ScienceResearch/BioinformaticsTools/ MicroarrayQualityControlProject/UCM134500.pdf.

For our analysis, we study the variance components of data under five one-color microarray platforms: Applied Biosystems (ABI), Affymetrix (AFX), Agilent Technologies (AG1), GE Healthcare (GEH), and Illumina (ILM). The four titration pools of two distinct RNA samples and two mixtures of the original samples are labeled A through D. They consist of 100% Universal Human Reference RNA (UHRR), 100% Human Brain Reference RNA (HBRR), 75% UHRR and 25% HBRR, and 25% UHRR and 75% HBRR, for samples A, B, C, and D, respectively. Each microarray provider used its own software to produce quantitative and qualitative RNA analysis. Microarray hybridizations that did not meet quality control criteria were removed from analysis. There remained between 56 and 60 arrays under each platform. In total, there were 293 arrays included in our analysis. We use the entire mapped set of 12 091 common genes across microarray platforms.

7.3.2 Normalization

In order to develop accurate statistical methodology for microarray expression data, a researcher first has to successfully remove systematic sources of variability. This is done through a process called normalization, which is designed to remove variability that exists for any reason other than biological differentiation in the samples of interest. Normalization of data allows for the comparison of expression levels across microarrays. For one-color arrays, we often find it necessary to correct the shape of gene intensity distributions so that they are similar between arrays. Common normalization methods include quantile normalization (Bolstad *et al.* 2003) and global median normalization (Zien *et al.* 2001; Quackenbush 2002).

Global median normalization forces the arrays to have equal median intensity. Data from the ith array are normalized by subtracting the median intensity from all intensity values from the same array. Specifically, the normalized data are given by

$$x_i^{norm} = x_i - median\{x_i\},$$

where x_i and $median\{x_i\}$ are the intensity values and the median intensity, respectively, of the ith array. Quantile normalization forces each array to have the same empirical distribution of intensity values. The quantile normalized data are given by

$$x_i^{norm} = F^{-1}(G_i(x_i)),$$

where G_i is estimated by the empirical distribution of the ith array and F is estimated by the empirical distribution of the average sample quantiles over all arrays.

We consider both normalization procedures for our analysis of the MAQC data. First, we use a variant of global median normalization to achieve consistency between arrays. The intensity values on each array were both centered and scaled. For each ith array, we subtract the median intensity from all intensity values and divide the difference by the standard deviation. After normalization, the median measurement of each array will be 0 and the standard deviation will be 1. Secondly, we apply quantile normalization to the standardized data to further reduce the variability observed between arrays. The standardized arrays are separated into 12 groups formed by the distinct site (1,2,3) by sample (A,B,C,D) factor combinations. We perform quantile normalization on the set of arrays within each of the 12 groups. The two sets of normalized data will help to determine whether the choice of normalization affects the comparability of microarray platforms.

In Figure 7.1, we present boxplots of the raw data, global median normalized data, and quantile normalized data for the first 15 arrays. It is evident that normalization aligns the discrepancy in central location that is seen in the distributions of raw data. Both

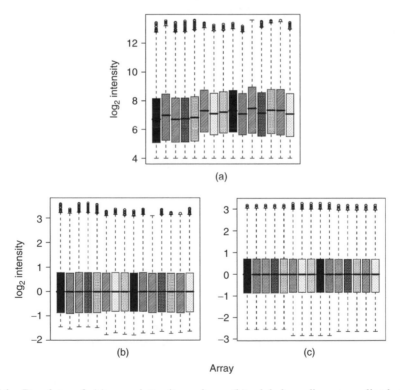

Figure 7.1 Boxplots of (a) raw intensity values, (b) global median normalized data, and (c) quantile normalized data for the first 15 microarrays.

methods of normalization transform the data so that the median intensity is the same across arrays. We also observe that quantile normalization of the standardized data corrects for positive skewness in each of the array data. Unlike global median normalization, quantile normalization adjusts the data so that the distribution of each array is more normal and bell-shaped.

7.3.3   Gene-Specific ANOVA Model

A mixed effects model was fitted to perform the variance component analysis. For each gene, measured expression levels, $y_{g,ijk}$, are modeled by

$$y_{g,ijk} = \mu_g + \alpha_{g,i} + b_{g,j} + \varepsilon_{g,ijk},$$

where μ_g is the overall mean expression, $\alpha_{g,i} \sim N(0, \sigma_s^2)$ is the random effect of site $i = 1, 2, 3$, $b_{g,j}$ is the fixed effect of sample $j = 1, \ldots, 4$, and $\varepsilon_{g,ijk} \sim N(0, \sigma_e^2)$ are independently and identically distributed error terms for i, j, and $k = 1, \ldots, 5$. The model was fitted separately on expression data for each of the five microarray platforms. Among all genes, the proportion of terms with significant interaction between the site and sample factors, $(\alpha b)_{g,ij}$, ranged between 3% and 19% for each platform. Thus, the interaction term is excluded from the model and we consider the reduced, additive model.

7.4   Application: The MAQC Project

We estimate the fixed sample effect of gene g over all five microarray platforms. Since sample is treated as a fixed variable in the mixed ANOVA model, we measure the effect based on samples A, B, C, and D separately for each gene. The fixed effect of the tth sample is the tth sample mean, computed as

$$\mu_{g,t} = \frac{1}{\# \text{ of observations in } t\text{th sample}} \sum_{\{j|j=t\}} y_{g,ijk}.$$

As a preliminary step, we evaluate how well the fixed effect estimates agree between microarray platforms. This is carried out for both normalization methods. Overall, the centers of the distribution of fixed effect estimates are equivocal across platforms for each sample. However, we find that quantile normalization of the raw data presents greater similarity in the variability and shape of the empirical distributions across platforms. Figures 7.2 and 7.3 show density plots of the fixed effect estimates for global median normalized data and quantile normalized data, respectively.

The degree of variability in the microarray platforms is assessed using the two-component ANOVA model. Variance components are estimated on a gene-by-gene basis. For each gene, total variation can be decomposed into the sum of site and error variance, i.e. $\sigma_y^2 = \sigma_s^2 + \sigma_e^2$. With one-color microarray platforms, the array and residual error variance components are confounded. Hence, σ_e^2 represents both sources of variation.

The sample average and standard error of estimated variance components across all genes are reported in Table 7.1 for each microarray platform. Within a given normalization procedure, these measures of variability are relatively consistent across the ABI, AFX, AG1, and ILM platforms. However, the average estimated site and error variances

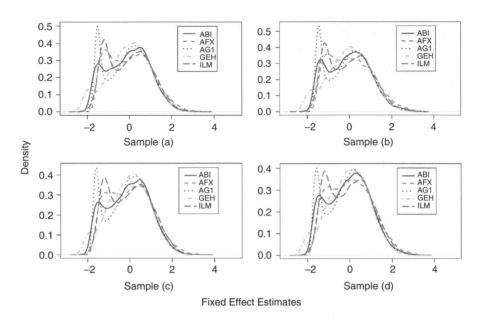

Figure 7.2 Density plots of the fixed effect estimates for gene data obtained from the five microarray platforms. A different graph is presented for each of the sample factors. Data were normalized using global median normalization. (Color version available as Figure S7.2 on website.)

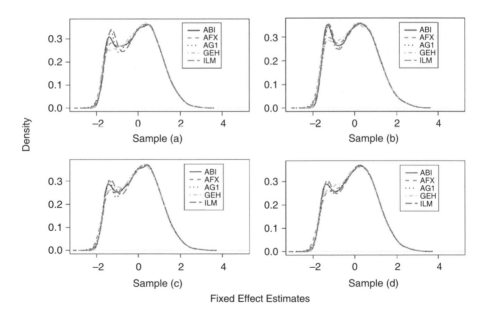

Figure 7.3 Density plots of the fixed effect estimates for gene data obtained from the five microarray platforms. A different graph is presented for each of the sample factors. Data were normalized using quantile normalization. (Color version available as Figure S7.3 on website.)

Table 7.1 Average (standard error) estimated variance of random effects from fitted mixed effects ANOVA model. Data were normalized using global median and quantile normalization.

Platform	Site		Error	
	Global median	Quantile	Global median	Quantile
ABI	0.009 (0.019)	0.008 (0.019)	0.022 (0.030)	0.021 (0.028)
AFX	0.006 (0.011)	0.006 (0.011)	0.005 (0.006)	0.007 (0.009)
AG1	0.011 (0.031)	0.011 (0.031)	0.009 (0.017)	0.011 (0.023)
GEH	0.041 (0.055)	0.038 (0.054)	0.048 (0.089)	0.026 (0.045)
ILM	0.008 (0.011)	0.006 (0.010)	0.013 (0.020)	0.015 (0.019)

are significantly higher in the GEH platform. Under global median normalization, the GEH average exceeds the average variability in other platforms by a factor that ranges between 3.7 and 6.8 for site variation and between 2.2 and 9.6 for error variation. This factor ranges between 3.5 and 6.3 for site variability under quantile normalization. We observe little difference in the summary estimates across normalization methods, with one exception – quantile normalization is shown to reduce the average error variability in the GEH platform by nearly half. It is also apparent from the averages presented in Table 7.1 that many of the platforms average reasonably small variability ($\sim$0.01), which is indicative of good intra-platform reproducibility.

To provide a more in-depth look at the agreement in variability across microarray platforms, we look at the absolute difference in variance estimates gene by gene for each pairwise combination of platforms. Here, we confirm previous findings which suggest a high level of concordance in variability between the ABI, AFX, AG1, and ILM platforms. Although the normalization method has no effect among comparisons between the methods, we again find that quantile normalization reduces the array and residual error variation for GEH data. In Table 7.2, we list the average absolute difference across genes for all 10 pairwise comparisons.

Table 7.2 Average absolute difference of variance estimates between two platforms. Data are reported for site and error variance components for data normalized by global median and quantile normalization.

Comparison	Site		Error	
	Global median	Quantile	Global median	Quantile
ABI–AFX	0.010	0.009	0.019	0.016
ABI–AG1	0.014	0.013	0.018	0.018
ABI–GEH	0.038	0.036	0.042	0.023
ABI–ILM	0.010	0.009	0.016	0.014
AFX–AG1	0.011	0.011	0.007	0.009
AFX–GEH	0.038	0.036	0.044	0.021
AFX–ILM	0.008	0.007	0.009	0.01
AG1–GEH	0.040	0.038	0.044	0.023
AG1–ILM	0.013	0.011	0.011	0.013
GEH–ILM	0.038	0.036	0.042	0.021

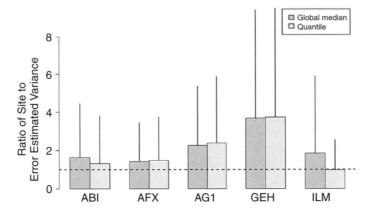

Figure 7.4 Barplot of mean with standard deviation bar for ratio values of site to error estimated variance. Data were normalized using global median and quantile normalization. The dashed horizontal line represents perfect equivalence of site to error estimated variance (i.e. $\hat{\sigma}_s^2/\hat{\sigma}_e^2 = 1$).

Table 7.3 Average ratio of site to error estimated variance. Data were normalized using global median and quantile normalization.

	Normalization method	
Platform	Global median	Quantile
ABI	1.639	1.313
AFX	1.422	1.480
AG1	2.263	2.390
GEH	3.684	3.732
ILM	1.843	0.980

Previous studies have shown that site variability plays an important role in microarray data analysis (Waring *et al.* 2004; Chen *et al.* 2007). Specifically, Chen *et al.* (2007) concluded that although the patterns of expression levels within each platform were similar, site-to-site variation was the main source of variability and contributed greatly to the observed differences between platforms. Our findings agree with previous arguments concerning the contribution of site-to-site variability. In Figure 7.4 we plot the mean with standard deviation bar of $\hat{\sigma}_s^2/\hat{\sigma}_e^2$ ratios across genes. Overall, site variability accounts for the majority of variation within a platform. This trend is particularly predominant in AG1 and GEH where the relative difference between site and error variability exceeds 2. The averages reported for ILM deviate most between the two methods of normalization. Otherwise, we observe minimal difference across normalization procedures (Table 7.3).

7.5   Discussion and Conclusion

Several different commercial microarray platforms are widely used in large-scale gene expression experiments. Preferably, a researcher could analyze data from one platform and be confident in his or her ability to reproduce analogous results using another complementary platform. However, there are fundamental differences in the processes used to generate data. Each platform differs in its protocol for RNA isolation, probe preparation, and synthesis and target labeling, all of which affect the measured intensity of a gene expression. Thus, the comparison and amalgamation of data from different platforms and/or laboratory sites are challenging.

The data analysis presented here centers on the level of variability agreement between gene expression data generated by five one-color microarray platforms. In addition to assessing the variability between platforms, we also analyze the concordance between global median normalization and quantile normalization. Our research indicates strong agreement in variability across the ABI, AFX, AG1, and ILM platforms for both normalization methods. On many assessment levels, we observe greater variability in measurements obtained from the GEH platform. However, quantile normalization of the data significantly reduces variability in the random error variance component. While the choice of normalization method does not affect site-to-site variation, it does play a role in the observed technical error variability between microarray platforms. For example, quantile normalization is shown to do a better job in aligning the empirical distributions of fixed effect estimates across platforms. This suggests that better accuracy can be obtained through quantile normalization. Our findings also suggest that much of the variability in gene expression data is due, in large part, to variability between laboratory sites.

Acknowledgements

The authors would like to thank Naisyin Wang for her helpful comments.

8

Variance due to Smooth Bias in Rat Liver and Kidney Baseline Gene Expression in a Large Multi-laboratory Data Set

Michael J Boedigheimer, Jeff W Chou, J Christopher Corton, Jennifer Fostel, Raegan O'Lone, P Scott Pine, John Quackenbush, Karol L Thompson, and Russell D Wolfinger

Abstract

To characterize variability in baseline gene expression, the ILSI Health and Environmental Sciences Institute Technical Committee on the Application of Genomics in Mechanism Based Risk Assessment recently compiled a large data set from 536 Affymetrix arrays for rat liver and kidney samples from control groups in toxicogenomics studies. Institution was one of the prominent sources of variability, which could be due to differences in how studies were performed or to systematic biases in the signal data. To assess the contribution of smooth bias to variance in the baseline expression data set, the robust multi-array average data were further processed by applying loess normalization and the degree of smooth bias within a data set was characterized. Bias correction did not have a large effect on the results of analyses of the major sources of variance but did affect the identification of genes associated with certain study factors if significant smooth bias was present within the data set.

8.1   Introduction

Gene expression analysis using whole genome microarrays is increasingly being used as an additional tool with traditional toxicity testing studies on pharmaceuticals and

Batch Effects and Noise in Microarray Experiments: Sources and Solutions edited by A. Scherer
© 2009 John Wiley & Sons, Ltd

environmental toxicants for deriving predictive and mechanistic insights. Toxicogenomics studies are typically highly variable in design, because dose delivery conditions need to be tailored to the physiochemical and pharmacological characteristics of a drug in order to mimic human exposure routes or to maximize levels of exposure in animal studies. Although other study parameters, such as time of day of sacrifice and withholding food for 24 hours before sacrifice (i.e. fasting), are known to impact the expression levels of sets of genes (Almon *et al.* 2008; Kita *et al.* 2002), this level of study protocol information is often not provided in published and other publicly available data sets.

The ILSI Health and Environmental Sciences Institute (HESI) Technical Committee on the Application of Genomics in Mechanism Based Risk Assessment recently compiled a freely accessible data set of control animal microarray data (Boedigheimer *et al.* 2008). Although control animal data were recognized to be a potential resource for analysis of baseline fluctuations in gene expression due to biological or technical factors, until now control animal data had not been publicly available on a scale and in a form suitable for data mining. This data set consists of Affymetrix array data contributed by HESI participants in the US and Europe on rat liver and kidney samples from animals that received vehicle alone or were untreated within control groups in toxicogenomics studies. For each sample, information was collected on common variables in toxicity studies (e.g. dosing regimen) and other factors that have known confounding effects on toxicity studies (strain, supplier, gender, diet, and age) (Kacew 2001).

The baseline expression data set consists of signal data from 536 microarrays from 16 institutions and 48 in-life studies. The data set includes data collected on three Affymetrix rat expression array types, RGU34A ($n = 192$), RAE230A ($n = 213$), and RAE230_2 ($n = 131$), for two tissues, liver ($n = 396$) and kidney ($n = 140$). The data set also included three rat strains, Sprague Dawley ($n = 302$), Wistar ($n = 210$), and Fischer 344/N ($n = 24$), and both males ($n = 436$) and females ($n = 100$). Details on the distribution of 22 of the 38 study factors within the collected data are available online (Boedigheimer *et al.* 2008). The data set was filtered for outliers using data from Affymetrix report files generated using the MAS 5.0 algorithm. Seventeen scans exceeded arbitrary cutoffs for scaling factor (>10), average background (>120), or maximum background (>130) and were excluded from further analysis. One data set that was identified as having been generated using a high photomultiplier tube (PMT) setting was also removed from the collection. PMT setting was one of the major sources of variability among the Affymetrix array data generated in the multi-laboratory toxicogenomics project conducted by the HESI Genomics Committee (ILSI Health and Environmental Sciences Institute 2003). Although earlier models of scanners for Affymetrix arrays allowed for adjustment of PMT settings (high or low), the newer models have a fixed gain, so this source of variability is only applicable to comparisons of archival data with more recently generated data.

The 483 quality-filtered files of array data were processed using robust multi-array average (RMA) separately for each of the three array types in the data set and analyzed using multivariate statistical and graphical techniques for the contribution of different study factors to baseline variability in gene expression. A subset of the 35 biological and technical factors obtained for each sample was evaluated in detail for their contribution to variability in expression. Gender, organ section (for kidney), strain, and fasting state (primarily for liver) were the study factors that emerged as key sources of variability. The data set was also used to identify genes which had the most and least inherent variability, were gender-selective, or altered by fasting.

A few institutions used multiple laboratories and most laboratories performed multiple studies. Laboratory was a prominent source of variability in most tissue–array sets. The observed contribution to variance by laboratory could be due to differences in how the studies were performed or to systematic biases in the signal data. The aforementioned analyses were performed without correcting for potential batch effects. Differences in strain, fasting status, and gender cause large effects on gene expression that were easily identifiable sources of variance in the baseline expression data set. However, more subtle changes in expression that could be contributed by differences in diet or vehicle, for example, may not be detectable in the data sets without correcting for smooth bias.

8.2 Methodology

RMA normalization was performed as previously described (Boedigheimer *et al.* 2008). Loess normalization was computed using JMP Genomics version 3.2, which uses SAS/STAT PROC LOESS. The normalization was performed separately for data from each tissue–array combination. An overall average of $\log_2$ RMA-normalized data from each group was used as the baseline for estimating the smooth bias curve. A smoothing parameter of 0.2 was used, representing a window size of 20% of the data. Principal components were computed for each data set using Matlab version R2007a. In this analysis components were computed for each sample after centering, but not scaling the expression values for each reporter. The first two components were plotted in a scatterplot using different shapes and shades for selected study factors.

To determine the presence or absence of a signal, $\log_2$ intensities of perfect match probes were compared to mismatch probes from the same probe set (feature) within an array using a two-way analysis of variance (ANOVA). An alpha value of 0.05 was used as a threshold to make a present call.

Logistic regression of present calls on RMA signal intensity data was performed using the binomial distribution and a logit link function.

8.3 Results

8.3.1 Assessment of Smooth Bias in Baseline Expression Data Sets

The difference in $\log_2$ transformed RMA data between an individual sample and the group average produces a $\log_2$ ratio of intensity that is plotted against the average intensity to create ratio–intensity (RI) plots (also known as *MA* plots) which reveal a distinct smooth bias in most of the baseline expression data. We investigated the impact of this smooth bias by removing it with loess normalization. Loess normalization estimates the smooth bias curve using a weighted local regression and subtracts it from the data. Normalizing data like these using both RMA and loess is an attempt to delicately remove unwanted technical sources of variability while retaining other sources of interest, a form of data 'surgery'. The process is imperfect because of the complex confounding patterns in these data, but aims to provide some assurance that final sources of interest are not due to technical artifacts.

The rat baseline expression data set contains control liver data from 359 samples, 17 laboratories, 42 studies, and on three different Affymetrix array types (RAE230A, RAE230_2, and RGU34A). Data sets of different tissue–array combinations were evaluated for smooth bias as follows: for each probe set, we calculated the mean $\log_2$ signal value across the set of liver-array data and the deviation from the mean for that probe set in a given sample. The extent of smooth bias can be visualized in RI plots of the deviation from the mean versus the mean signal value (see Figure 8.1(a)). These plots also demonstrate that loess normalization effectively removed smooth bias in the baseline expression data set (see Figure 8.1(b)).

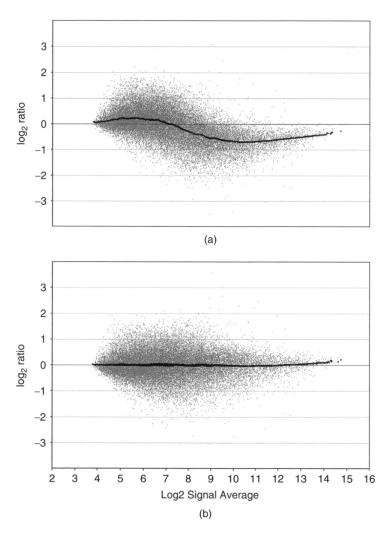

Figure 8.1 RI plots for one lab-study 06A sample (#3–3) compared to the average for the liver-RAE230_2 data set using RMA normalized data (a) before and (b) after loess normalization. Small gray dots are $\log_2$ ratio values for individual probe sets, black dots are mean $\log_2$ ratio values.

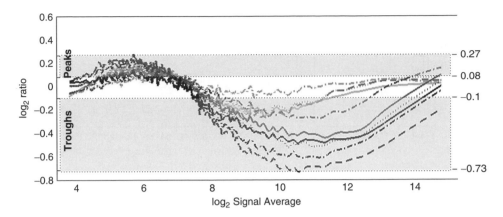

Figure 8.2 Mean log2 values from RI plots for the ten samples in lab-study 06A from the liver-RAE230_2 data set.

The smooth bias within a data set varies in amplitude, but is consistently sinusoidal, as shown in Figure 8.2 with lab-study 06A samples in the liver-RAE230_2 data set. For different samples, the peak and trough amplitudes in the smooth bias occurred either below or above an average $\log_2$ signal value around 7 depending on the direction of bias relative to the mean of the data set. This crossover point ($y = 0$) coincides with the median data set value. A high positive bias in low intensity range represents an average compression in signal at a specific intensity range in one array relative to another. This implies that arrays with such bias will have less dynamic range or sensitivity to detect low expression transcripts.

The range between peak and trough amplitudes for each sample is referred to as the MaxDiff value. Since smooth bias is assessed relative to average signals across a tissue–array data set, the MaxDiff values can be used to compare the extent of smooth bias between array types, laboratories, and studies. Of the three rat expression Affymetrix array formats in the data set, the format with the smaller feature size (RAE230_2) tended to have more smooth bias within and between laboratories than the other sets (Figure 8.3). For example, the ten liver samples in study A from lab 6 in the liver-RAE230_2 set showed the largest interquartile range in MaxDiff values (Figures 8.2 and 8.3).

8.3.2 Relationship between Smooth Bias and Signal Detection

A potential underlying explanation for the smooth bias is variation in signal to noise ratios between arrays. This might arise, for example, due to high background on an array. A graphical analysis (Figure 8.4) reveals a strong association between fraction of probe sets detected on an array (i.e. present calls) and the signal intensity near the first peak in smooth bias (Figure 8.2), which occurs around an RMA normalized $\log_2$ intensity of 5.6 for these arrays. A deeper examination of the detection rate was performed using a logistic regression for the two arrays that showed the greatest difference in direction of bias in this region. Using this technique, the fraction of present calls was estimated along the

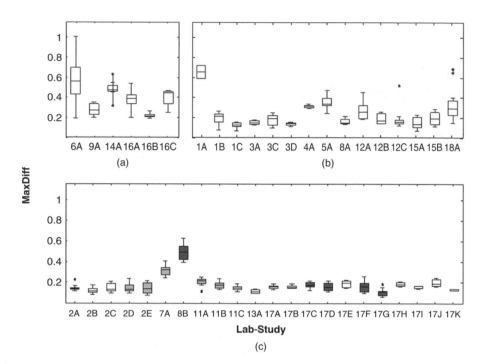

Figure 8.3 Box plots of MaxDiff values for samples in each lab-study for (a) the liver-RAE230_2, (b) liver-RGU34A, and (c) liver-RAE230A tissue–array data sets. Boxes represent the interquartile range with upper and lower whisker limits; lines within boxes indicate the median; crosses indicate outliers. Liver-RAE230A sets used to identify genes differentially expressed by fasting are indicated in (c) by color. Dark gray indicates sets that were fasted, light gray indicates sets that were non-fasted, and open boxes were not used in the analysis.

entire intensity spectrum (Figure 8.4(b)). The plots show that the most positively biased array had a lower detection rate throughout the curve and had a slightly slower transition in the middle intensity than the most negatively biased array. This implies that a given transcript may be below the limits of detection on one array but within the working range of another array. Counterintuitively the signal intensity of a transcript may be higher on an array where it is undetected than on an array where it is detected as present.

8.3.3 Effect of Smooth Bias Correction on Principal Components Analysis

Prominent sources of variability in baseline expression data were assessed using several multivariate techniques that were applied separately on the different tissue–array sets for 17 of the study factors in Boedigheimer *et al.* (2008). In the original study, principal components analysis showed that organ section in control kidney samples was a prominent source of variability. The pattern of gene expression in kidney medulla showed distinct separation from whole kidney or kidney cortex gene expression. In most

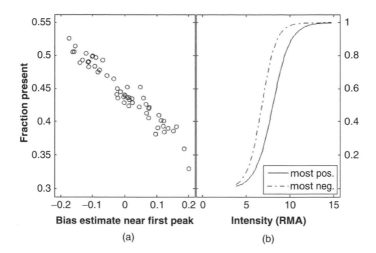

Figure 8.4 High bias near first peak of the RI plots is associated with lower signal detection rates. (a) Scatterplot of fraction of probe sets detected as present on the RAE230_2 arrays versus the array bias at RMA signal of 5.6. (b) Plot of predicted fraction present by intensity for the two arrays with the most extreme bias near the first peak.

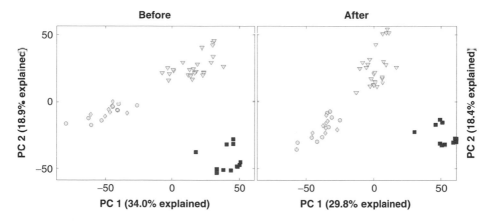

Figure 8.5 Principal component analysis reveals small differences after loess normalization. The first two principal components are plotted for control liver samples run on RAE230_2 arrays for data before and after loess normalization. Marker shapes indicate laboratories and shading represents fasting status.

other tissue–array set comparisons, laboratory and certain study factors (e.g. fasting) were prominent explanatory factors for sample clustering in principal component scatterplots. To assess the effect of smooth bias correction on these variability analyses, principal component scores were computed for each tissue–array combination before and after loess normalization. Figure 8.5 shows the results for control rat liver samples run on RAE230_2 arrays. The overall structure of the data does not appear to be greatly influenced by loess normalization, but there are noticeable differences in scale and rotation of the first two components. Similar results were seen in the five other tissue–array combinations.

8.3.4   Effect of Smooth Bias Correction on Estimates of Attributable Variability

We recomputed the Hotelling–Lawley and variance component statistics from Boedigheimer *et al.* (2008), designed to assess proportional contribution of known factors to total variability in the data. In the original analysis, the study factors that had the largest and most reproducible contributions to variance included gender, strain, fasting, organ section, and laboratory. Loess adjustment caused a reduction in some of the Hotelling–Lawley components, for example, Lab and Study in liver-RGU34A and Gender and Vehicle in liver-RAE230A (see Table 8.1). The variance component estimates

Table 8.1 Proportion of variability attributable to observed factors before and after loess adjustment for smooth bias for the three liver data sets.

| | Liver-RGU34A ($n = 124$) | | | | Liver-RAE230A ($n = 179$) | | | | Liver-RAE230_2 ($n = 56$) | | | |
| | HL | | VC | | HL | | VC | | HL | | VC | |
Factor	−L	+L	−L	+L	−L	+L	−L	+L	−L	+L	−L	+L
Lab	**0.48**	0.34	0.06	**0.21**	0.26	0.28	0.00	0.00	**0.73**	**0.62**	0.00	0.00
Study	**0.42**	0.04	0.02	0.03	0.22	0.10	0.02	0.01	0.07	0.06	0.02	0.02
Gender	0.19	0.06	**0.21**	**0.17**	**0.46**	0.22	**0.50**	**0.46**	0.00	0.00	0.00	0.00
Organ Section	0.19	0.09	0.02	0.04	0.21	0.21	0.00	0.00	**0.73**	**0.62**	0.00	0.00
Strain	0.09	0.03	0.01	0.01	0.19	0.13	**0.06**	**0.06**	**0.71**	**0.47**	0.04	0.03
Fasted	0.11	0.04	**0.09**	**0.20**	0.17	0.08	**0.17**	**0.17**	**0.66**	**0.40**	**0.14**	**0.16**
Sac Method	0.11	0.06	0.02	0.00	0.24	0.16	**0.06**	0.00	**0.71**	**0.47**	0.00	0.00
Anesthetic	0.02	0.01	0.07	0.01	0.12	0.06	0.00	0.00	0.15	0.09	0.00	0.00
Age	0.03	0.02	0.00	0.02	0.05	0.03	0.00	0.00	0.15	0.11	0.05	0.05
Route	0.09	0.06	**0.13**	**0.12**	0.04	0.03	0.00	0.00	0.06	0.04	0.01	0.03
Dose Freq	0.03	0.02	0.05	0.01	0.01	0.00	**0.04**	**0.05**	0.06	0.04	0.01	0.01
Fixation	0.10	0.04	0.06	0.02	0.03	0.02	**0.02**	0.02	0.00	0.00	0.00	0.00
Diet	0.12	0.05	0.01	0.03	0.17	0.10	0.01	0.02	**0.69**	**0.54**	**0.46**	**0.40**
Vehicle	0.10	0.04	0.02	0.02	**0.35**	0.04	0.02	**0.05**	**0.73**	**0.53**	0.05	0.01
RNA Amount	0.03	0.02	0.03	0.02	0.18	0.15	**0.07**	**0.11**	**0.73**	**0.62**	0.00	0.00
Dose Duration	0.04	0.03	0.01	0.02	0.23	0.14	0.00	0.00	**0.97**	**0.93**	**0.12**	**0.18**
Scanner	0.16	0.08	**0.09**	0.01	0.18	0.09	0.01	0.02	0.00	0.00	0.00	0.01
Total Var	641	610			854	817			3430	2870		
Full model	70	203			166	358			245	427		
Residual			0.08	0.07			0.02	0.03			0.10	0.10

Note: HL = Hotelling–Lawley, which estimates proportion of variability after adjusting for all other effects. VC = variance components, which is a simultaneous partitioning of total variability. L (−) without or (+) with loess normalization. Bold numbers indicate values deemed significantly large, as in Boedigheimer *et al.* (2008).

did not change appreciatively. Additional information on confounding relationships is available in supplementary data available online (see Boedigheimer *et al.* 2008).

8.3.5   Effect of Smooth Bias Correction on Detection of Genes Differentially Expressed by Fasting

In preclinical toxicity studies, animals are often fasted just prior to study completion in order to enhance assessments of liver histopathology, although fasting has a profound effect on liver gene expression. As a test of the utility of the baseline expression data collected from multiple laboratories to identify known biological differences, an analysis of genes significantly changed by fasting samples was performed across 115 samples in the liver-RAE230A control expression data set (Boedigheimer *et al.* 2008). The degree of smooth bias seen in these samples is depicted in Figure 8.3(c), with the largest effect seen for laboratory 8. The effect of smooth bias correction was assessed on detection of gene expression profiles in the livers of animals that had either been fasted or nonfasted. The resultant fasting gene lists were compared to published results for liver from Wistar rats that had been fasted for 24 hours (Nakai *et al.* 2008). Loess normalization resulted in a reduction in the total number of genes that were called significant (from 444 to 388) but an increase in the number of genes called significant from the Nakai *et al.* study (from 982 to 1007). A comparison of the overlap in the two data sets is shown in Figure 8.6. From

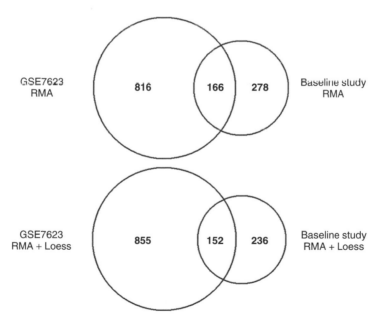

Figure 8.6   Overlap in the fasting genes identified in the baseline data set and in the Nakai *et al.* (2008) study (GSE7623, Gene Expression Omnibus http://www.ncbi.nlm.nih.gov/geo/). Gene sets were identified after RMA normalization alone or with loess normalization. Overlaps in the gene sets between the studies are shown in the Venn diagrams.

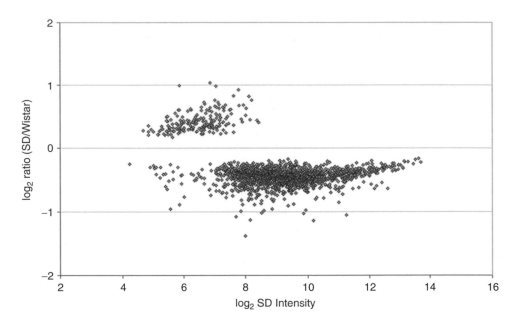

Figure 8.7   RI plots of strain-selective probe sets detected as significantly differentially expressed only before applying loess normalization.

the baseline expression study using RMA alone there was a 37% overlap with the Nakai *et al.* study. Loess normalization increased the overlap only incrementally (to 39%). Thus the results indicate that loess normalization did not appreciably change the identification of fasting genes in these data.

8.3.6   Effect of Smooth Bias Correction on the Detection of Strain-Selective Gene Expression

Of the three rat expression array types in the data set, the set of RAE230_2 array data had the highest level of smooth bias within and between laboratories (Figure 8.3(a)). We examined the effect of correcting for smooth bias within this set on the identification of genes associated with strain differences. Strain was shown to be one of the significant contributors to variance in baseline expression in our previous analysis (Boedigheimer *et al.* 2008). The liver-RAE230_2 set contains data from laboratory 14 on male, non-fasted Sprague Dawley rats, from laboratory 09 on male, fasted F344/N rats, and from laboratories 06 and 16 on male, nonfasted Wistar rats. Fasting has a strong effect on gene expression (Boedigheimer *et al.* 2008), so samples from fasted animals (all F344/N studies) were excluded from this analysis. Wistar and Sprague Dawley are both outbred stocks of laboratory rat that are commonly used in toxicology studies. This analysis has some limitations because, although it is controlled for fasting and gender, there are differences in parameters other than strain between laboratories that could also contribute to differences in gene expression levels.

Gene transcript levels that were significantly different between Sprague Dawley and Wistar rats were identified using the two-sample t-test (with random variance model) in BRB-ArrayTools version 3.5.0 developed by Richard Simon and Amy Peng Lam, with a p-value threshold of 0.001 (see Wright and Simon 2003). A similar number of gene transcripts were identified as differentially expressed without loess normalization (5797) and after loess normalization (5330). Most of the genes that were identified as differentially expressed genes between strains were detected with either data transformation method (3863). However, there was a clear directional bias and an uneven distribution in intensity in the strain-selective gene changes detected in the non-loess normalized data (1729 higher and 4068 lower in Sprague Dawley relative to Wistar). After loess normalization, the strain-selective genes showed a more balanced distribution (2621 higher and 2709 lower in Sprague Dawley relative to Wistar). Visualization on RI plots of the genes that were identified as strain-selective before but not after loess normalization of the data indicates that these data have a shape similar to the sinusoidal smooth bias curve in Figure 8.1 for laboratory 06 samples (Figure 8.7).

8.4   Discussion

Although methods for normalization of individual data sets are well established, methods for comparison of data between studies are still not well established. Dealing effectively with the various sources of bias in data is essential to make effective use of the available information and to allow for comparison of data across laboratories and between studies. This is particularly important in applications of toxicogenomics, where organizations such as the Food and Drug Administration are working to incorporate genomic assays into the approval process used in evaluating compound safety and efficacy. But the implications of improved protocols that could allow broad data comparison go far beyond toxicogenomics and have implications in virtually every field of biomedical inquiry.

The recent analysis of a large collection of arrays that came from control animals in toxicogenomics experiments (Boedigheimer *et al.* 2008) provided an opportunity to address questions about the comparability of gene expression data generated at different laboratories or different times. In that study, where information was collected about many common experimental factors, it was shown that biological factors known to induce gene expression changes were significant sources of variation despite being collected at different times and places. Since that study did not attempt to adjust for systematic bias, further work was done to examine what methods might be used for visualizing bias in the data and what effect bias adjustment would have on the results. The data set was gathered from 48 different studies from 16 institutions and therefore represents a reasonable sampling of the variability one might expect to encounter when comparing different data sets.

To detect bias in these data we used a graphical approach in which the difference between the group mean of a gene from its overall mean is plotted against its overall mean (RI plot). We found that a locally weighted scatterplot smoothing (loess) curve through these data is a highly sensitive technique for visualizing this type of bias and allows for simultaneous visualization of data from multiple arrays. We also found a striking negative association between the amount of bias in the low intensity regions and the number (or fraction) of probe sets detected on an array (Figure 8.4). A high bias in this region implied not just an overall lower detection rate but a lower detection rate throughout most of the signal intensity range. This suggests that for affected probe sets signal intensity is unrelated, or at least not highly related, to transcript abundance. Future work may reveal how to best deal with these probe sets.

In the baseline expression data set, the highest degree of smooth bias within and between laboratories was observed in the control rat liver data generated on Affymetrix RAE230_2 arrays. This array format has a smaller feature size that requires the use of new labeling and hybridization protocols than were used for earlier generations of array formats. In the baseline data set, the higher level of smooth bias observed in RAE230_2 data could represent a lower degree of operator experience with the optimal conditions for generating consistent data with this newer array format.

The smooth bias can be effectively removed using loess normalization, a method that has a long history in the normalization of two-color microarray data. This correction did not have a large effect on analyses of the major sources contributing to variance in the baseline expression data set. A principal components analysis revealed only small changes in the amount of variance explained in the first few components and in the overall structure. In addition, a variance components analysis showed that the overall partitioning of variance was relatively unchanged. A second approach, using the Hotelling–Lawley trace, showed some hints that the data had less variability overall and that more of it was explainable after loess normalization. In analyses of genes with significant differences in expression level for a given study factor, loess normalization tended to have the greatest effect on detection of genes that were changed to a modest degree (less than twofold) and for data sets with the greatest degree of identified smooth bias.

Based on this study it seems that examining data for smooth bias is a prudent course to take when analyzing data generated at different times or laboratories. Plotting significant findings in the context of the bias can reveal potential artifacts that may be removed by loess correction. If the bias is small, loess correction is likely to yield modest reductions in batch effects and may allow a greater degree of variability to be attributable to experimental sources. If the bias is large, it implies there may be a qualitative difference in the assays. In this case probe sets that are near the detection limits should be subjected to special scrutiny. In the cases we studied, the known sources of biological variability were still manifest despite visible systematic bias, demonstrating both the potential and limitations of cross-study analysis of microarray data.

Although our work represents an important step in dealing with systematic effects in comparison of gene expression between various sources, additional work remains. One of the most important questions is which genes should be included in the loess normalization. Ideally, one would like to identify a set of genes that, in a particular data set, is not likely to be differentially expressed – essentially the equivalent of a set of housekeeping genes. While array experiments have largely done away with the idea that there are genes that are universally invariant in their expression levels, it is likely that in each particular tissue there is a core set of genes which are, on average, expressed at comparable levels; rank-invariant normalization methods are based on this assumption and one that we intend to explore further.

Finally, despite the fact that the work presented here has focused on gene expression measures obtained using DNA microarrays, the general results here should in no way be viewed as being limited to microarrays. Although new approaches, such as 'digital gene expression' measurements using next generation DNA sequencing technologies, promise to eliminate some of the systematic sources of bias observed in microarray data, they most certainly will introduce their own particular biases and biological effects will continue to confound comparisons of data between laboratories and across studies. Following normalization, the methods we have presented are agnostic as to the underlying technology used to generate the data and, we believe, will at least serve as a starting point for analysis of expression data derived using other technologies.

Acknowledgements

The contribution of JF to this work was supported by the Division of Intramural Research of the National Institute of Environmental Health Science, under contract HHSN273200700046U. The HESI data set is available from CEBS (http://cebs.niehs. nih.gov/) under accession number 008-00003-0001-000-2.

9

Microarray Gene Expression: The Effects of Varying Certain Measurement Conditions

Walter Liggett, Jean Lozach, Anne Bergstrom Lucas, Ron L Peterson, Marc L Salit, Danielle Thierry-Mieg, Jean Thierry-Mieg, and Russell D Wolfinger

Abstract

This chapter explores measurements from an experiment with a batch effect induced by switching the mass of RNA analyzed from 400 ng to 200 ng. The experiment has as additional factors the RNA material (liver, kidney, and two mixtures) and the RNA source (six different animals). We show that normalization can partially correct the batch effect. On the basis of the normalized intensities, we compare, gene by gene, the size of the batch effect with the animal-to-animal variation. These comparisons show that the animal variation is larger or smaller depending on which gene is considered. We present gene-by-gene tests of the linearity of the microarray response. In addition, we present data analysis results that suggest other batch effects.

9.1 Introduction

In metrological terminology, conditions of measurement can be held constant, or certain conditions can be allowed to vary (ISO, 2007). Measurements made under constant conditions (also referred to as repeatability conditions) constitute a batch. A set of measurements may include several batches that are delineated by changes in certain conditions. Variations observed among the batches beyond what is observed under constant conditions are batch effects. In this paper, we illustrate batch effects in microarray measurements by examining variations related to two measurement conditions. The measurements were

Batch Effects and Noise in Microarray Experiments: Sources and Solutions edited by A. Scherer
© 2009 John Wiley & Sons, Ltd

made with the Agilent Whole Rat Genome Microarray 4×44 K*. The two measurement conditions allowed to vary are the amplification and labeling input mass and the substrate on which four microarrays are mounted.

In contrast to the familiar metrology of scalar measurands, batch effects in microarray gene expression have many facets. The microarray considered here has 45 018 probes, and, consequently, technical variation manifests itself in 45 018 dimensions. For this reason, there are many choices to be made in the characterization of microarray batch effects. On the one hand, one does not want to present a characterization that is incomprehensible because it is overwhelming in its detail. On the other hand, one does not want to hide manifestations of the batch effects that might be helpful in improving measurement protocols or methods for statistical inference. The purpose of this chapter is to present a characterization that strikes a balance. We illustrate the various ways that batch effects show up in measurements and describe methods for analyzing batch effects. These ideas can be applied in other situations.

The structure of the experiment considered here provides many possibilities for data analysis. There are measurements on liver RNA and kidney RNA, which are very different, and on mixtures of these RNAs to allow insight into calibration-curve linearity. Measurements were made on RNAs from six animals (*Rattus norvegicus*). Because these animals formed a control group for a previous study, their RNAs were expected to be similar. In addition, technical replicates are included. Guo *et al.* (2006) performed an experiment with six animals as biological replicates but without mixtures or technical replicates.

The data set considered here consists of 96 microarray measurements made on RNAs from six animals (R. norvegicus). RNA was extracted from both the liver and the kidney of each animal. Two mixtures of the liver and kidney RNAs were prepared, one with 3 parts liver and 1 part kidney and the other with 1 part liver and 3 parts kidney. Pure liver, the two mixtures and pure kidney from each animal were each measured in triplicate leading to 72 measurements. In order of decreasing liver fraction, the liver, the two mixtures and the kidney are labeled A, B, C, and D. Liver RNAs from the six animals were pooled; the kidney RNAs were similarly pooled; and $3 + 1$ and $1 + 3$ mixtures of these pools were prepared. These four materials were each measured in quadruplicate leading to 16 measurements. In addition, the liver RNAs from the first three animals were pooled; the kidney RNAs were similarly pooled; mixtures were prepared; and the four materials were measured without replication. Finally, the RNAs from the second three animals were pooled, mixed and measured. Files with complete data details are available through the companion website at www.the-batch-effect-book.org. These details might suggest additional analyses. Affymetrix measurements of the same 96 materials have been reported in Liggett *et al.* (2008).

This paper is organized as successive steps in uncovering the input mass effects, which are more pronounced than the substrate effects in our experiment. It should be noted that the two levels of input mass are completely confounded with choice of operator and with some other potential causes of batch effects. As is typical in microarray data analysis, we begin with normalization (Stafford 2008). Our normalization method makes use of the mixture relations among the RNAs, which has as one result a partial correction of batch effects (Liggett 2008).

---

* Certain commercial equipment, instruments, or materials are identified in this paper to foster understanding. Such identification does not imply recommendation or endorsement by the National Institute of Standards and Technology, nor does it imply that the materials or equipment identified are necessarily the best available for the purpose.

For each probe, we model the normalized intensities with a linear mixed model (Pinheiro and Bates 2000). Such a model involves fixed effects such as the animal effects and the input mass effects, random effects such as the substrate effect, and the random error. Finally, we consider evidence of unrecorded changes in measurement conditions.

9.2   Input Mass Effect on the Amount of Normalization Applied

In this section and the next, we present an analysis of the measurements on the individual animals, which consist of results from 72 arrays. These measurements are among the 96 that can be downloaded from ArrayExpress (E-TABM-555). ArrayExpress gives results from each microarray as a spreadsheet. To allow our analysis to be repeated, we give its starting point in terms of the headings of the five spreadsheet columns we use. Our analysis is based on background corrected intensities given by gBGSubSignal divided by gMultDetrendSignal. We eliminate intensities from control probes (ControlType) and from probes that exhibit feature nonuniformity (glsFeatNonunifOL) or saturation (glsSaturated) for any of the 72 individual-animal arrays. Thus, for each array, intensities from 42 697 probes of the 45 018 enter our analysis.

Generally, the observed intensities from a group of arrays exhibit inter-array differences that must be dealt with either by a pre-processing step referred to as normalization or by some other means. We normalize the 12 arrays from each individual animal separately using a two-step process (Liggett 2008). The first step is array-by-array global normalization of the intensities. For each array, we compute a scale factor by first adding 30 to each background-corrected intensity because some intensities are reported as negative, and then by finding the geometric mean of these values. We then divide each intensity by this scale factor. The second step is adjustment of the globally normalized intensities so that the linear model implied by the mixing of the materials fits better. Letting the globally normalized intensity for array i of animal j and for probe g be y_{jig}, the adjusted intensity is $(y_{jig} - \eta_{0ji})/\eta_{ji}$. The values of η_{0ji} and η_{ji} are computed to improve the fit of the adjusted intensities to the linear model specified by the mixing.

In addition to the estimates of η_{0ji} and η_{ji}, our two-step normalization process gives a fitted model for the intensities. As implied by the mixing, the intensities can be modeled as $x_{Aji}\theta_{Ajg} + x_{Dji}\theta_{Djg}$. The fraction of the liver material in the mixture x_{Aji} is 1 for material A, $3/(3 + \varphi_j)$ for material B, $1/(1 + 3\varphi_j)$ for material C, and 0 for material D. The symbol φ_j denotes the ratio of the concentration of mRNA in the kidney RNA to the concentration of mRNA in the liver RNA. We have $x_{Dji} = 1 - x_{Aji}$. Liggett et al. (2008) describe estimation of φ_j in the second step of the normalization. We use the estimates of φ_j, θ_{Ajg}, and θ_{Djg} in specifying the analysis in the next section.

Figure 9.1 shows the estimated normalization parameters for each of the six animals. The points are labeled with the user designation A or B, which corresponds to input mass of 200 ng and 400 ng. We see that the second step of our normalization process generally scales the 200 ng arrays up in comparison to the 400 ng arrays. This can be viewed as our normalization providing a partial correction of the input mass batch effect.

9.3   Probe-by-Probe Modeling of the Input Mass Effect

For each probe, there are 72 normalized intensities that are differentiated by animal, material (liver, kidney or a mixture), input mass, and substrate. By fitting a linear mixed

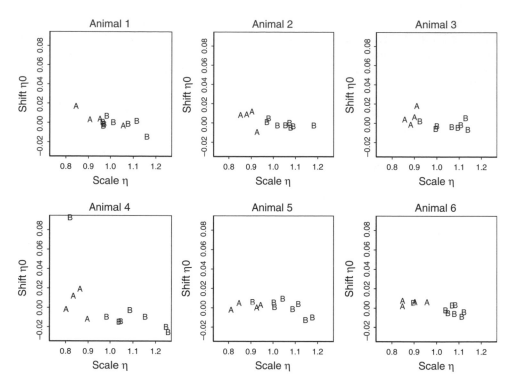

Figure 9.1   Normalization parameters for each array with input mass 200 ng (A) and 400 ng (B). Normalization consists of subtraction of the shift followed by division by the scale.

model (Pinheiro and Bates 2000) to these intensities, we can distinguish the contributions of these factors. Of particular interest is comparison of the animal-to-animal variation with the input mass effect. In case–control studies, the central statistical question is whether the amount of animal variation within the control group can account for the apparent differences between the cases and the controls. If the input mass effect is of the same size as or larger than the amount of animal variation, then this effect has the potential for misleading researchers in the formulation of study conclusions.

It is, of course, not enough to compare animal variation with input mass effect for each probe. Summarization over the probes is necessary in the understanding of the input mass effect. In our approach, we confine the summarization to a subset of the probes and choose a model parameterization with coefficients that are comparable from one probe to another. One aspect of selecting the probe subset is choice of probes for which there is appreciable liver mRNA or appreciable kidney mRNA. Another aspect is selecting a probe subset for which there is a model parameterization appropriate for summarization.

In our approach, the model parameterization is based on representing the animal-to-animal variation and the input mass effect in terms of fractional deviations from the animal average intensities. Such an approach has the advantage of making the probes with strongest response comparable regardless of their liver–kidney difference or over-all intensity response. The average intensities are obtained from the intensity estimates obtained in the normalization process. These are animal-by-animal estimates of the liver intensity $\hat{\theta}_{Ajg}$ and the kidney intensity $\hat{\theta}_{Djg}$. We denote the average of the liver intensities by $\bar{\theta}_{Ag}$ and the average of the kidney intensities by $\bar{\theta}_{Dg}$. For array i of animal j, the animal average intensity is given by

$$\hat{x}_{Aji}\bar{\theta}_{Ag} + \hat{x}_{Dji}\bar{\theta}_{Dg} = (\bar{\theta}_{Ag} + \bar{\theta}_{Dg})/2 + (\hat{x}_{Aji} - \hat{x}_{Dji})(\bar{\theta}_{Ag} - \bar{\theta}_{Dg})/2.$$

In our notation, we apply a 'hat' to $\hat{x}_{Aji}$ and $\hat{x}_{Dji}$ because they incorporate the estimate of φ_j obtained as part of the normalization. In modeling, we parameterize the animal effects with an intercept term $\alpha_{jg}(\bar{\theta}_{Ag} + \bar{\theta}_{Dg})/2$ and with a slope term $\beta_{jg}(\hat{x}_{Aji} - \hat{x}_{Dji})(\bar{\theta}_{Ag} - \bar{\theta}_{Dg})/2$. Similarly, we parameterize the input mass effect with the terms $\gamma_g(\bar{\theta}_{Ag} + \bar{\theta}_{Dg})/2$ and $\delta_g(\hat{x}_{Aji} - \hat{x}_{Dji})(\bar{\theta}_{Ag} - \bar{\theta}_{Dg})/2$.

The probes we select for summarization satisfy two conditions. First, the probes satisfy

$$((\bar{\theta}_{Ag} > 3) \text{ AND } (\bar{\theta}_{Dg} > 0)) \text{ OR } ((\bar{\theta}_{Ag} > 0) \text{ AND } (\bar{\theta}_{Dg} > 3)).$$

In general terms, this condition amounts to the requirement that either the liver mRNA or the kidney mRNA have appreciable intensity. As a way of gauging this condition, we note that there are 9620 probes for which $\bar{\theta}_{Ag} > 3$ and 11 107 probes for which $\bar{\theta}_{Dg} > 3$. Second, the probes must exhibit a fold change of at least 2 between the liver and the kidney. In other words, the probes satisfy

$$\left(\left(\frac{\bar{\theta}_{Ag}}{\bar{\theta}_{Dg}}\right) > 2\right) \text{ OR } \left(\left(\frac{\bar{\theta}_{Dg}}{\bar{\theta}_{Ag}}\right) > 2\right).$$

These two conditions ensure that the model parameterization we have adopted is not misleading. The number of probes that satisfy these two conditions is 5014.

With the substrate effect left out, the modeling equation for the normalized intensity u_{jig} corresponding to array i for animal j is given by

$$u_{jig} = (\alpha_{jg} + \gamma_{ig})(\bar{\theta}_{Ag} + \bar{\theta}_{Dg})/2 + (\beta_{jg} + \delta_{ig})(\hat{x}_{Aji} - \hat{x}_{Dji})(\bar{\theta}_{Ag} - \bar{\theta}_{Dg})/2 + \varepsilon_{jig},$$

where

$$\text{Var}(\varepsilon_{jig}) = \sigma_g^2\left[\left(\max\left(\hat{x}_{Aji}\hat{\theta}_{Ajg} + \hat{x}_{Dji}\hat{\theta}_{Djg}, 0\right)\right)^2 + \omega^2\right].$$

In this modeling equation, the unknown parameters are $\alpha_{jg}, \beta_{jg}, \gamma_{ig}, \delta_{ig}$ and σ_g^2. The other quantities are obtained from the normalization results. The parameters γ_{ig} and δ_{ig} are restricted to $\pm\gamma_g$ and $\pm\delta_g$ because the input mass has only two levels. We note that the estimates of the animal effect parameters α_{ji} and β_{ji} and the estimates of the batch effect parameters γ_{ig} and δ_{ig} are essentially uncorrelated because of the design of the experiment.

In Figure 9.2, the input mass effect is shown as a plot of the intercept coefficient γ_g versus the slope coefficient δ_g for the selected probes. The points lie close to the $x = y$ line. This means that most of the input mass effect shows up as a fractional deviation from the average intensity $\delta_g(\hat{x}_{Aji}\bar{\theta}_{Ag} + \hat{x}_{Dji}\bar{\theta}_{Dg})$, that is, as a multiplicative effect on the average intensity. Relative to this, the remainder of the input mass effect $(\gamma_g - \delta_g)(\bar{\theta}_{Ag} + \bar{\theta}_{Dg})/2$ is small. This leads us to thinking about the modeling results in terms of the fractional deviation from the intensity δ_g and the remaining fractional deviation $\gamma_g - \delta_g$.

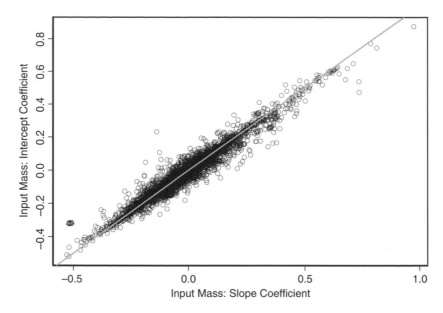

Figure 9.2   The input mass effect for 5014 selected probes. The effect is given as fractional change in the calibration-curve intercept and fractional change in the calibration-curve slope.

Consider comparison of the input mass effect with the animal variation. On the basis of what is shown in Figure 9.2, we begin by comparing $\delta_g(\hat{x}_{Aji}\bar{\theta}_{Ag} + \hat{x}_{Dji}\bar{\theta}_{Dg})$ with the animal-by-animal changes in $\beta_{jg}(\hat{x}_{Aji}\bar{\theta}_{Ag} + \hat{x}_{Dji}\bar{\theta}_{Dg})$. To make this comparison, we compute, for each probe, the standard deviation of β_{jg}, $j = 1, \ldots, 6$, the fractional deviations of the animal intensities. For the group of probes selected, the top two panels of Figure 9.3 show histograms for the fractional deviations from the average intensities. The top panel shows the input mass effect. The middle panel shows the standard deviation of the six animal values. We see that these two histograms have roughly the same spread. To be more exact in the comparison of the two histograms, we must take into account that the animal variation is given as a single standard deviation although two standard deviations would come closer to covering all the animal variation. Moreover, the probe-to-probe variation in the animal standard deviation also involves sampling error. In other words, the standard deviations shown are spread out in part by the estimation error inherent in a sample of only six animals and in part by actual probe-to-probe variation in the standard deviation. A more exact comparison seems to require that we introduce a more definite context.

There is more to the animal variation than $\beta_{jg}(\hat{x}_{Aji}\bar{\theta}_{Ag} + \hat{x}_{Dji}\bar{\theta}_{Dg})$, the fractional deviation from the average intensity. There is $(\alpha_{jg} - \beta_{jg})(\bar{\theta}_{Ag} + \bar{\theta}_{Dg})/2$. The bottom panel of Figure 9.3 shows a histogram of the standard deviation of $\alpha_{jg} - \beta_{jg}$, the remaining fractional deviation. Comparison of the middle and bottom panels of Figure 9.3 shows that the remaining fractional deviation is generally smaller.

Another batch effect might be expected on the basis of the microarray mounting, which entails sets of four arrays with each set mounted on a common substrate. We model the substrate effect as a random effect given by $\varsigma(\hat{x}_{Aji}\bar{\theta}_{Ag} + \hat{x}_{Dji}\bar{\theta}_{Dg})$, where ς varies

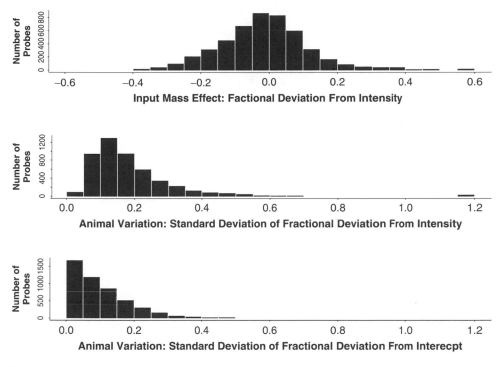

Figure 9.3   The input mass effect compared to the animal-to-animal variation and the animal intensity variation compared to the animal intercept variation. Histograms for 5014 selected probes.

randomly from the arrays on one substrate to the arrays on another. We can perform a likelihood ratio test of the null hypothesis that there is no substrate effect. Were exact p-values for this test available, we could combine the results from a set of probes using false discovery rate methodology (Storey and Tibshirani 2003). Because this is not the case (Pinheiro and Bates 2000), the current situation is somewhat more complicated.

Figure 9.4 shows the estimate of the standard deviation of ς plotted versus the likelihood ratio test statistic for the 5014 probes selected. Values of the test statistic below 0.001 have been replaced by 0.001. Points for most of the probes lie in the lower right-hand corner where the test statistic and the standard deviation are both nearly 0. We see that few of the probes have substrate-effect standard deviations that are comparable to the animal variation shown in Figure 9.3. Moreover, only 588 of the probes have values for the likelihood ratio test statistic greater than 1.

On the basis of theoretical results, it is generally believed that the likelihood ratio test statistic under the null hypothesis is distributed as a mixture of two distributions, a probability mass at 0 and a chi-square distribution with 1 degree of freedom. The probability mass at 0 can be seen in the probes shown in the lower left of Figure 9.4. The mixing proportion for the distributions must generally be determined by simulation (Pinheiro and Bates 2000).

Applying these ideas, we conclude that the substrate effect should be of little concern. We have derived a mixing proportion from the observed distribution of the likelihood

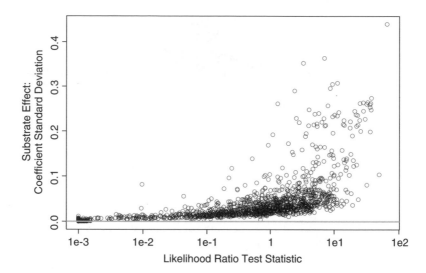

Figure 9.4 Two characterizations of the substrate effect for 5014 probes: an estimate of the standard deviation of the substrate effect on the vertical axis, and a test statistic for the hypothesis of no substrate effect on the horizontal axis. The 3706 probes with likelihood ratio test statistic less than 0.001, which display essentially no evidence of a substrate effect, are plotted at the lower end of the horizontal axis.

ratio test statistic. The resulting mixture distribution fits the observed distribution closely. Moreover, the random effect standard deviations are relatively small compared to the animal variation.

9.4   Further Evidence of Batch Effects

The choice of 200 ng of material for some of the Aglient microarray measurements was not originally part of the experimental design. Rather, two (microarray) users were assigned the task of making the 96 measurements, 72 on individual animal RNAs and the rest on RNAs made from pooling the individual animals RNAs. The user designated A in the ArrayExpress description of the measurements made a mistake in performing the 48 measurements assigned. In remaking these measurements, only 200 ng of material was generally available for each RNA sample. In that the 200 ng measurements were a recovery from a mistake, it would not be surprising if other batch effects showed up. The samples obtained by pooling the liver RNAs of the six animals and the kidney RNAs provide an opportunity to investigate this. From these pools, 3:1 and 1:3 mixtures were made as with the individual animal RNAs. The pooled materials were then measured in quadruplicate. User A made all these measurements.

Consider using the measurements on the six-animal pools for testing the linearity of the relation between concentration and the intensity. We begin with the same normalization procedure that we applied to the measurements on the individual animal RNAs. The normalized intensities are given by u_{pig}. Linearity implies that

$$u_{pig} = \hat{x}_{Api}\theta_{Apg} + \hat{x}_{Dpi}\theta_{Dpg} + \varepsilon_{pig},$$

where the variance of ε_{pig} is given by

$$\mathrm{Var}(\varepsilon_{pig}) = \sigma_g^2 \left[\left(\max \left(\hat{x}_{Api} \hat{\theta}_{Apg} + \hat{x}_{Dpi} \hat{\theta}_{Dpg}, 0 \right) \right)^2 + \omega^2 \right].$$

The parameters θ_{Apg}, θ_{Dpg}, and σ_g^2 are unknown, and values for the other parameters are taken from the output of the normalization. If the realizations of ε_{pig} are independent from measurement to measurement, then we can estimate σ_g^2 from the four sets of replicates regardless of the material-to-material relations among the intensities. This estimate provides the denominator for an F test. Otherwise, there are four average intensities, one for each material, and a model of the relation among these averages with two parameters θ_{Apg} and θ_{Dpg}. This gives another estimate of σ_g^2. The ratio of the two estimates of σ_g^2 can be used as statistic for a lack-of-fit test. Under the null hypothesis, the test statistic is an F ratio with 2 and 12 degrees of freedom.

We consider all the probes for which

$$\left(\left(\hat{\theta}_{Apg} > 3 \right) \text{ AND } \left(\hat{\theta}_{Dpg} > 0 \right) \right) \text{ OR } \left(\left(\hat{\theta}_{Apg} > 0 \right) \text{ AND } \left(\hat{\theta}_{Dpg} > 3 \right) \right)$$

except for the control probes and those showing nonuniformity or saturation. There are 11 530 such probes. Figure 9.5 shows a quantile–quantile plot for the log of the F ratio. If the null hypothesis were satisfied for every probe, the curve would fall on the $x = y$ line. We see that the curve is above this line and that there is further deviation from this line at the upper end. The deviation at the upper end is evidence that some probes exhibit saturation. Similar behavior has been seen in other expression microarray data. The separation of the curve from the $x = y$ line over the whole range is more puzzling.

That the curve in Figure 9.5 is above the $x = y$ line could be evidence of an unsuspected batch effect. That the F ratio is too large could be the result of the denominator of the F ratio being too small. Supposedly, the denominator is computed from four independent replicate measurements on each material. If there were an unsuspected batch effect, the independence assumption would not be valid. The F test is based on the hypothesis that the observed variation within the sets of replicates accounts for the observed deviation from a linear calibration curve. If there is batch structure that largely coincides with the replicate sets, then this hypothesis will not be true and the denominator of the F ratio will be too small.

An investigation of an unknown batch effect on the basis of the 16 measurements from one probe would have limited power. Combining measurements from all the selected probes offers more possibilities for insight. Let the residuals from the material means be denoted r_{pig}. Consider the standardized version of these residuals

$$r_{pig} / \sqrt{\left(\max \left(\hat{x}_{Api} \hat{\theta}_{Apg} + \hat{x}_{Dpi} \hat{\theta}_{Dpg}, 0 \right) \right)^2 + \omega^2}.$$

Regarding each probe as a replicate, we can compute a 16×16 covariance matrix from these standardized residuals. This covariance matrix might suggest the form of the unknown batch effect. One way to proceed is to perform a principal components analysis on this covariance matrix. Such an analysis suggests that the replicate-to-replicate variation for materials A and B (more liver RNA) is larger than that for materials C and D (more kidney RNA). This is consonant with the results shown in Figure 9.5, although the cause of the unknown batch effect is still not clear.

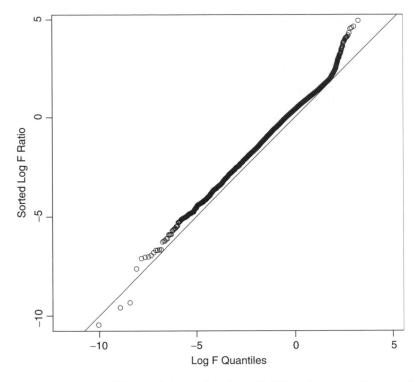

Figure 9.5 Measurements of the pooled materials from 11 530 probes: quantile–quantile plot of the F test for linearity of the calibration curve.

9.5 Conclusions

Examination of some particular batch effects suggests aspects of a general approach to dealing with batch effects. Not addressed in this chapter is the initial step of identifying potential batch effects. This chapter started with potential batch effects already identified. These came from the list of effects such as user (operator) effects and interlaboratory effects that are usually considered in metrology. Potential effects are often identified through remeasurement of reference materials such as the materials used in the MicroArray Quality Control project (MAQC Consortium 2006). Because microarrays provide a multivariate response, clues can also come from single-array quality measures (Brettschneider *et al.* 2008).

Once potential batch effects have been identified, the methods illustrated in this chapter can be applied. Note that user was identified at the outset and before the 200 ng–400 ng protocol difference arose. The design and data analysis methods of the current study seem appropriate for use in the study of other batch effects. Particularly appropriate is the probe-by-probe fitting of linear mixed models with terms either fixed or random for each batch effect. These methods have two advantages. First, the inclusion of liver, kidney, and mixture RNAs allows the batch effects to be portrayed in terms of slopes of calibration curves. Second, the inclusion of several animals allows comparison of biological variation with technical variation.

The batch effect caused by the change in input mass cannot be completely erased through data analysis. This shows that one cannot rely on data analysis to solve problems with batch effects. One solution is to reduce the batch effect through protocol modification. For example, stricter adherence to 400 ng as the input mass, which is the mass specified in the initial measurement protocol, would have been preferable to relying on the data analysis. Because complete elimination of a batch effect may not be possible, a complementary solution is choice of experimental design. For example, the data analysis methods presented here were able to separate the animal variation from the user effect because the design specified the operator factor as orthogonal to the animal and material factors. In particular, for the individual animal measurements, one and only one of every three replicate measurements was made by user A with the 200 ng protocol. Were one user to have measured four animals and the other user two animals, the measurements would have been much harder to interpret.

10

Adjusting Batch Effects in Microarray Experiments with Small Sample Size Using Empirical Bayes Methods

W Evan Johnson and Cheng Li

Abstract

Nonbiological experimental variation or batch effects are commonly observed across multiple batches of microarray experiments, often rendering the task of combining data from these batches difficult. The ability to combine microarray data sets is advantageous to researchers to increase statistical power to detect biological phenomena from studies where logistical considerations restrict sample size or in studies that require the sequential hybridization of arrays. In general, it is inappropriate to combine data sets without adjusting for batch effects. Methods have been proposed to filter batch effects from data, but these are often complicated and require large batch sizes (at least 25) to implement. Because the majority of microarray studies are conducted using much smaller sample sizes, existing methods are not sufficient. We propose parametric and nonparametric empirical Bayes frameworks for adjusting data for batch effects that are robust to outliers in small sample sizes and perform comparably to existing methods for large samples.

10.1 Introduction

With the many applications of gene expression microarrays, biologists are able to efficiently extract hypotheses that can later be tested experimentally in a laboratory setting.

This chapter is based on the article by Johnson, WE, Li, C, & Rabinovic, A, 2007, 'Adjusting batch effects in microarray expression data using empirical Bayes methods', *Biostatistics*, vol. 8, pp. 118–127, originally published by Oxford University Press.

For example, a microarray experiment might compare the gene expression profile of diseased or treated tissue (treatment) with the profile of normal tissue (controls) to determine which genes are associated with the disease or the presence of the treatment, providing better understanding of disease–gene relationships. However, practical considerations limit the number of samples that can be amplified and hybridized at one time, and replicate samples may be generated several days or months apart, introducing systematic batch effects or nonbiological differences that make samples in different batches not directly comparable. Batch effects have been observed from the earliest microarray experiments (Lander 1999), and can be caused by many factors, including the batch of amplification reagent used, the time of day when an assay is done, or even the atmospheric ozone level (Fare *et al.* 2003; see also Chapter 2 in the present volume). Batch effects are also inevitable when new samples or replicates are incrementally added to an existing array data set or in a meta-analysis of multiple studies that pools microarray data across different labs, array types or platforms (Rhodes *et al.* 2004).

Combining experiments results in better utilization of existing data, so biologists can combine their own data from multiple experiments, or others can conduct meta-analytic research using multiple published microarray data sets. However, batch effects often account for a large portion of the variation in two or more combined experiments, preventing the pooling of replicates to enhance power to detect real gene changes, generate new biological hypotheses, and making it difficult to distinguish other biological effects from the data. Some researchers have presented methods for adjusting for batch effects (Alter *et al.* 2000, Benito *et al.* 2004, Shabalin *et al.* 2008), but these methods require many samples (at least 25) in each batch for best performance and may remove real biological variation from the data. In this chapter we develop an empirical Bayes (EB) method that is robust for adjusting for batch effects in data whose batch sizes are small.

10.1.1 Bayesian and Empirical Bayes Applications in Microarrays

Bayesian and empirical Bayes (EB) methods have been applied to a large variety of settings in microarray data analysis. Bayesian methods are very appealing in microarray problems because of their ability to robustly handle high-dimensional data when sample sizes are small. Bayesian methods are primarily designed to 'borrow information' across genes and experimental conditions in hope that the borrowed information will lead to better estimates or more stable inferences.

EB models have been widely applied to analyze microarray data (Chen *et al.* 1997; Efron *et al.* 2001; Kendziorski *et al.* 2003; Gottardo *et al.* 2005; Lönnstedt *et al.* 2005; Pan 2005). In particular, Newton *et al.* (2001) derived an EB framework to model the distributions of expression ratios for cDNA arrays using the gamma and lognormal distributions. This method was designed to stabilize the expression ratios for genes with very high or very low ratios, possibly protecting their inference from outliers in the data. Tusher *et al.* (2001) presented a method of detecting differential expression (SAM analysis) that can be considered an early attempt at an EB adjustment for probe variances, by using a modified t statistic where the variance estimate is slightly inflated by some constant (same for all genes). Later, Smyth (2004) used EB to stabilize variances across genes using an EB variance estimate derived by shrinking variances across all other genes. In this chapter we extend the EB methods to the problem of adjusting for batch effects in microarray data.

10.2   Existing Methods for Adjusting Batch Effect

10.2.1   Microarray Data Normalization

Microarray data are often subject to high variability due to noise and artifacts, often attributed to differences in chips, samples, labels, etc. In order to correct these biases caused by nonbiological conditions, researchers have developed normalization methods to adjust data for these effects. For example, Schadt *et al.* (2001) and Tseng *et al.* (2001) use sets of rank invariant (nondifferentially expressed) genes or probes to normalize arrays to prepare the data for analysis. For oligonucleotide arrays, this method is designed to eliminate systematic array effects by adjusting the overall array brightness to be similar across chips before computing expression values. Other normalization methods have also been developed (Bolstad *et al.* 2003; Yang *et al.* 2002c).

However, normalization procedures do not adjust the data for batch effects. For example, suppose that one set of genes within one batch of data have unusually high expression due to a batch effect. When this batch is normalized with another batch, the chip-to-chip brightness will be the same across batches, but the expression values will still be significantly higher in the first batch. Therefore, when combining batches of data (particularly batches that contain large batch-to-batch variation), normalization is not sufficient for adjusting for batch effects and other procedures must be applied.

10.2.2   Batch Effect Adjustment Methods for Large Sample Size

One of the first methods for adjusting data for batch effects is proposed in Alter *et al.* (2000). This paper is commonly known for its application of principle components analysis to the classification of microarray samples. However, the authors also present a method for adjusting arrays for batch effects, by adjusting 'the data by filtering out those eigengenes (and eigenarrays) that are inferred to represent noise or experimental artifacts'. This can be done by finding the singular value decomposition (SVD), recognizing the eigenvectors that represent batch effects, setting the 'batch effect' eigenvalues to zero, and reconstructing the data. Nielsen *et al.* (2002) successfully apply SVD batch effect adjustment to a microarray meta-analysis.

There are difficulties faced by researchers who try to implement SVD batch adjustment methods. For example, it is not always clear how to find the batch effect vector, and often it does not exist at all. Because the eigenvectors in the SVD are all orthogonal to each other the method is highly dependent on proper selection of the first several eigenvectors. A more direct approach is to use linear discriminant analysis (LDA), which no one has suggested in the literature, but this method (along with the SVD method) often breaks down in high-dimensional data when sample sizes are small. In addition, SVD and LDA approaches factor out all variation in the given direction, which may not be completely due to batch effects.

Benito *et al.* (2004) use distance weighted discrimination (DWD) to correct for systematic biases across microarray batches. DWD is an improvement on the support vector machine (SVM) for high-dimensional data when sample sizes are small. The SVM separates the data by maximizing the minimum distance from the data to a separating hyperplane, while DWD allows all the data (not just the data on the margin) to have influence on the determining the best hyperplane. Once the SVM or DWD hyperplane is found, the data are adjusted by projecting the different batches on the DWD plane, finding the batch mean, and then subtracting out the DWD plane multiplied by this mean. For best performance, DWD requires at least 25–30 samples per batch.

A more recent approach to combining multiple studies was presented by Shabalin *et al.* (2008). They applied an iterative clustering algorithm combined with and a block linear model to normalize arrays from different platforms. The clustering algorithm separates the gene sets into subgroups that appear co-expressed. Then a linear model is applied to normalize the arrays and remove batch effects. As with the previous methods, a large sample size is required for this approach.

Walker *et al.* (2008) applied a method for adjusting for batch effects using reference samples and the empirical Bayes methods presented in Chapter 11 in the present volume.

10.2.3   Model-Based Location and Scale Adjustments

Location and scale (LS) adjustments consist of a wide family of adjustments in which one assumes a model for the location (mean) and/or scale (variance) of the data within batches and then adjusts the batches to meet assumed model specifications. Therefore, LS batch adjustments assume that the batch effects can be modeled out by standardizing means and variances across batches. These adjustments can range from simple gene-wise mean and variance standardization to complex linear or nonlinear adjustments across the genes.

One straightforward LS batch adjustment is to mean-center and standardize the variance of each batch for each gene independently. This procedure typically results in a much-improved clustering heatmap of the data, which is very useful for researchers interested in classification by clustering. However, this procedure cannot be used for estimating gene fold changes because the loss of absolute magnitude of the gene expression values often results in nonsensical fold change estimates. Additionally, in more complex situations such as unbalanced designs or when incorporating numerical covariates, a more general LS framework must be used. For example, let Y_{ijg} represent the expression value for gene g for sample j from batch i. Define an LS model that assumes

$$Y_{ijg} = \alpha_g + \mathbf{X}\boldsymbol{\beta}_g + \gamma_{ig} + \delta_{ig}\varepsilon_{ijg} \tag{10.1}$$

where α_g is the overall gene expression, $\mathbf{X}$ is a design matrix for sample conditions, and $\boldsymbol{\beta}_g$ is the vector of regression coefficients corresponding to $\mathbf{X}$. The error terms, ε_{ijg}, can be assumed to follow a normal distribution with expected value of zero and variance σ_g^2. γ_{ig} and δ_{ig} represent the additive and multiplicative batch effects of batch i for gene g, respectively. The batch adjusted data, Y_{ijg}^*, is given by

$$Y_{ijg}^* = \frac{Y_{ijg} - \hat{\alpha}_g - \mathbf{X}\hat{\boldsymbol{\beta}}_g - \hat{\gamma}_{ig}}{\hat{\delta}_{ig}} + \hat{\alpha}_g + \mathbf{X}\hat{\boldsymbol{\beta}}_g, \tag{10.2}$$

where $\widehat{\alpha}_g, \widehat{\beta}_g, \widehat{\gamma}_{ig}, \widehat{\delta}_{ig}$ are estimators for the parameters $\alpha_g, \beta_g, \gamma_{ig}$, and δ_{ig} based on the model.

One major disadvantage of LS methods, as well as with SVD and DWD based methods, is that large sample (batch) sizes are required for implementation because such methods are dependent on the batch mean and variance estimates and are not robust to outliers in small sample sizes. However, in a typical microarray experiment it is common practice to minimize time and monetary constraints, commonly resulting in very few samples per experiment (less than 10). Additionally, the proposed LS methods only consist of gene-wise adjustments and do not consider systematic biases shared across genes.

10.3 Empirical Bayes Method for Adjusting Batch Effect

In this section we propose a method that robustly adjusts batches with small sample sizes. This method was developed to retain some of the flexibility of the gene-wise LS adjustments, while being robust to outlying measurements under small sample sizes. More importantly, this method incorporates systematic batch biases common across genes in making adjustments, assuming that phenomena resulting in batch effects often affect many genes in similar ways (increased expression, higher variability, etc.). Specifically, we estimate the LS model parameters that represent the batch effects by pooling information across genes in each batch to shrink the batch effect parameter estimates toward the overall mean of the batch effect estimates (across genes). These EB estimates are then used to adjust the data for batch effects, providing more robust (and possibly more accurate) adjustments for the batch effect on each gene. The method is described in three steps below.

10.3.1 Parametric Shrinkage Adjustment

We assume that the data have been normalized and expression values have been estimated for all genes and samples. We also filter out the genes which are 'Absent' in more than 80% of samples to eliminate noise. Suppose the data contain m batches containing n_i samples within batch i for $i = 1, \ldots, m$, for gene $g = 1, \ldots, G$. We assume the model specified in equation (10.1), and that the errors, ε, are normally distributed with mean zero and variance σ_g^2.

Step 1: Standardize the data
The magnitude of expression values could differ across genes due to mRNA expression level and probe sensitivity. In relation to equation (10.1), this implies that $\alpha_g, \beta_g, \gamma_g$ and σ_g^2 differ across genes, and if not accounted for, these differences will bias the EB estimates of the prior distribution of batch effect and reduce the amount of systematic batch information that can be borrowed across genes.

To avoid this phenomenon, we first standardize the data gene-wise so that genes have similar overall mean and variance. We estimate the model parameters $\alpha_g, \beta_g, \gamma_{ig}$ as $\widehat{\alpha}_g, \widehat{\beta}_g, \widehat{\gamma}_{ig}$ for $i = 1, \ldots, m$ and $g = 1, \ldots, G$.

For our examples from data sets 1 and 2, we use a gene-wise ordinary least squares approach to do this, constraining $\sum_i n_i \widehat{\gamma}_{ig} = 0$ (for all $g = 1, \ldots, G$) to ensure the identifiability of the parameters. We then estimate

$$\widehat{\sigma}_g^2 = \frac{1}{N} \sum_{ij} (Y_{ijg} - \widehat{\alpha}_g - \mathbf{X}\widehat{\boldsymbol{\beta}}_g - \widehat{\gamma}_{ig})^2$$

where N is the total number of samples. The standardized data, Z_{ijg}, are now calculated by

$$Z_{ijg} = \frac{Y_{ijg} - \widehat{\alpha}_g - \mathbf{X}\widehat{\boldsymbol{\beta}}_g}{\widehat{\sigma}_g}.$$

Step 2: EB batch effect parameter estimates using parametric empirical priors

We assume that the standardized data, Z_{ijg}, satisfy the distributional form

$$Z_{ijg} \sim N(\gamma_{ig}, \delta_{ig}^2).$$

Notice that the γ parameters here are not the same as in equation (10.1) – they are the γ from equation (10.1) divided by $\widehat{\sigma}_g$. Additionally, we assume the parametric forms for prior distributions on the batch effect parameters to be

$$\gamma_{ig} \sim N(\gamma_i, \tau_i^2) \quad \text{and} \quad \delta_{ig}^2 \sim \text{InverseGamma}(\lambda_i, \theta_i).$$

We apply Bayes' theorem to find the conditional (posterior) distribution of γ_{ig}, denoted $\pi(\gamma_{ig}|\mathbf{Z}_{ig}, \delta_{ig}^2)$, which satisfies

$$\pi(\gamma_{ig}|\mathbf{Z}_{ig}, \delta_{ig}^2) \propto L(\mathbf{Z}_{ig}|\gamma_{ig}, \delta_{ig}^2)\pi(\gamma_{ig})$$

$$\propto \exp\left\{-\frac{1}{2\delta_{ig}^2}\sum_j (Z_{ijg} - \gamma_{ig})^2\right\} \exp\left\{-\frac{1}{2\tau_i^2}\frac{1}{2}(\gamma_{ig} - \gamma_i)^2\right\}$$

$$= \exp\left\{-\frac{1}{2\delta_{ig}^2}\left[\sum_j Z_{ijg}^2 - 2\sum_j Z_{ijg}\gamma_{ig} + n_i\gamma_{ig}^2\right] - \frac{1}{2\tau_i^2}\left[\gamma_{ig}^2 - 2\gamma_{ig}\gamma_i + \gamma_i^2\right]\right\}$$

$$\propto \exp\left\{-\frac{1}{2}\left(\frac{n_i\tau_i^2 + \delta_{ig}^2}{\delta_{ig}^2\tau_i^2}\right)\left[\gamma_{ig}^2 - 2\left(\frac{\tau_i^2\sum_j Z_{ijg} + \delta_{ig}^2\gamma_i}{n_i\tau_i^2 + \delta_{ig}^2}\right)\gamma_{ig}\right]\right\}.$$

By completing the square, the distribution above can be determined to be the kernel of a normal distribution with expected value

$$E[\gamma_{ig}|\mathbf{Z}_{ig}, \delta_{ig}^2] = \frac{\tau_i^2\sum_j Z_{ijg} + \delta_{ig}^2\gamma_i}{n_i\tau_i^2 + \delta_{ig}^2}$$

which, given $\widehat{\gamma}_{ig}, \widehat{\delta}_{ig}^2, \overline{\gamma}_i$, and $\overline{\tau}_i^2$ as defined below, can be estimated as

$$\gamma_{ig}^* = \widehat{E}[\gamma_{ig}|\mathbf{Z}_{ig}, \delta_{ig}^{2*}] = \frac{n_i\overline{\tau}_i^2\widehat{\gamma}_{ig} + \delta_{ig}^{2*}\overline{\gamma}_i}{n_i\overline{\tau}_i^2 + \delta_{ig}^{2*}}.$$

For the conditional posterior distribution of δ_{ig}^2, given γ_{ig} and InverseGamma(λ_i, θ_i) prior, we note that

$$\pi(\delta_{ig}^2|\mathbf{Z}_{ig}, \gamma_{ig}) \propto L(\mathbf{Z}_{ig}|\gamma_{ig}, \delta_{ig}^2)\pi(\delta_{ig}^2)$$

$$\propto (\delta_{ig}^2)^{-\frac{n_i}{2}} \exp\left\{-\frac{1}{2\delta_{ig}^2}\sum_j (Z_{ijg} - \gamma_{ig})^2\right\} (\delta_{ig}^2)^{-(\lambda_i+1)} \exp\left\{-\frac{\theta_i}{\delta_{ig}^2}\right\}$$

$$= (\delta_{ig}^2)^{-\left(\frac{n_i}{2}+\lambda_i\right)-1} \exp\left\{-\frac{\theta_i + \frac{1}{2}\sum_j (Z_{ijg} - \gamma_{ig})^2}{\delta_{ig}^2}\right\},$$

which can be identified as an inverse gamma distribution with expected value

$$E[\delta_{ig}^2|\mathbf{Z}_{ig}, \gamma_{ig}] = \frac{\theta_i + \frac{1}{2}\sum_j (Z_{ijg} - \gamma_{ig})^2}{\frac{n_i}{2} + \lambda_i - 1}.$$

Given the method of moments estimates for θ_i and λ_i from below, the expectation above can be estimated by

$$\delta_{ig}^{2*} = \widehat{E}[\delta_{ig}^2|\mathbf{Z}_{ig}, \gamma_{ig}^*] = \frac{\bar{\theta}_i + \frac{1}{2}\sum_j (Z_{ijg} - \gamma_{ig}^*)^2}{\frac{n_i}{2} + \bar{\lambda}_i - 1}.$$

The hyperparameters γ_i, τ_i^2, λ_i, θ_i are estimated empirically from standardized data using the method of moments. Letting

$$\widehat{\gamma}_{ig} = \frac{1}{n_i}\sum_j Z_{ijg}$$

(batch i sample mean for gene g), estimates of γ_i and τ_i^2 are given respectively by

$$\bar{\gamma}_i = \frac{1}{G}\sum_g \widehat{\gamma}_{ig} \quad \text{and} \quad \bar{\tau}_i^2 = \frac{1}{G-1}\sum_g (\widehat{\gamma}_{ig} - \bar{\gamma}_i)^2.$$

Additionally, letting

$$\widehat{\delta}_{ig}^2 = \frac{1}{n_i - 1}\sum_j (Z_{ijg} - \widehat{\gamma}_{ig})^2$$

(batch i sample variance for gene g) we calculate

$$\bar{V}_i = \frac{1}{G}\sum_g \widehat{\delta}_{ig}^2 \quad \text{and} \quad \bar{S}_i^2 = \frac{1}{G-1}\sum_g (\widehat{\delta}_{ig}^2 - \bar{V}_i)^2$$

(mean and variance of $\widehat{\delta}_{ig}^2$). Setting the sample moments $\bar{V}_i$, $\bar{S}_i^2$ equal to the theoretical moments of the inverse gamma distribution, namely $\theta_i/(\lambda_i - 1)$ (mean) and

$\theta_i^2/[(\lambda_i - 1)^2(\lambda_i - 2)]$ (variance), and then solving this system yields estimates for λ_i and θ_i as follows:

$$\bar{\lambda}_i = \frac{\bar{V}_i + 2\bar{S}_i^2}{\bar{S}_i^2}, \quad \bar{\theta}_i = \frac{\bar{V}_i^3 + \bar{V}_i\bar{S}_i^2}{\bar{S}_i^2}.$$

These prior distributions (normal, inverse gamma) were selected due to their conjugacy with the normal assumption for the standardized data.

Based on the distributional assumptions above, the EB estimates for batch effect parameters, γ_{ig} and δ_{ig}^2, are given respectively by the conditional posterior means

$$\gamma_{ig}^* = \frac{n_i\bar{\tau}_i^2\widehat{\gamma}_{ig} + \delta_{ig}^{2*}\bar{\gamma}_i}{n_i\bar{\tau}_i^2 + \delta_{ig}^{2*}}, \quad \delta_{ig}^{2*} = \frac{\bar{\theta}_i + \frac{1}{2}\sum_j (Z_{ijg} - \gamma_{ig}^*)^2}{\frac{n_j}{2} + \bar{\lambda}_i - 1}. \tag{10.3}$$

Notice that the estimate γ_{ig}^* depends on δ_{ig}^{2*} and vice versa. There are no closed-form solutions for these parameters, and therefore they must be found iteratively. Starting with a reasonable value for δ_{ig}^{2*} (say, $\hat{\delta}_{ig}^2$), calculate an estimate of γ_{ig}^*. Now use the newly found value of γ_{ig}^* to estimate δ_{ig}^{2*}. Iterate the previous steps until convergence. This can be shown to be a simple case of the EM algorithm (Dempster *et al.* 1977), and typically only a few iterations (up to 30) are necessary to achieve very accurate estimates for the EB batch adjustments.

Step 3: Adjust the data for batch effects
After calculating the adjusted batch effect estimators, γ_{ig}^* and δ_{ig}^{2*}, we now adjust the data. The EB batch adjusted data Y_{ijg}^* can be calculated in a similar way to equation (10.2), but using EB estimated batch effects:

$$Y_{ijg}^* = \frac{\widehat{\sigma}_g}{\delta_{ig}^*}(Z_{ijg} - \widehat{\gamma}_{ig}^*) + \widehat{\alpha}_g + \mathbf{X}\widehat{\boldsymbol{\beta}}_g.$$

10.3.2 Empirical Bayes Batch Effect Parameter Estimates using Nonparametric Empirical Priors

In certain cases, the parametric prior assumptions may not be suitable for a particular data set, leading to the need for more flexible options for the prior distributions, so we use a nonparametric empirical prior to accommodate these data. We assume that the data has been standardized as in Step 1 above, and that the standardized data, Z_{ijg}, satisfy the distributional form

$$Z_{ijg} \sim N(\gamma_{ig}, \delta_{ig}^2).$$

Also let

$$\widehat{\gamma}_{ig} = \frac{1}{n_i}\sum_j Z_{ijg}, \quad \widehat{\delta}_{ig}^2 = \frac{1}{n_i - 1}\sum_j (Z_{ijg} - \widehat{\gamma}_{ig})^2,$$

as in the previous subsection. We estimate the batch effect parameters γ_{ig} and δ_{ig}^2, using estimates of the posterior expectations of the batch effect parameters, denoted $E[\gamma_{ig}]$ and

$E[\delta_{ig}^2]$ respectively. Let $\mathbf{Z}_{ig}$ be a vector containing Z_{ijg} for $j = 1, \ldots, n_i$. Given the posterior distribution $\pi(\mathbf{Z}_{ig}, \gamma_{ig}, \delta_{ig}^2)$ of the data $\mathbf{Z}_{ig}$ and batch effect parameters $\gamma_{ig}, \delta_{ig}^2$, the posterior expectation of γ_{ig} is given by

$$E[\gamma_{ig}] = \int \gamma_{ig} \pi(\mathbf{Z}_{ig}, \gamma_{ig}, \delta_{ig}^2) \, d(\gamma_{ig}, \delta_{ig}^2).$$

Now let $\pi(\gamma_{ig}, \delta_{ig}^2)$ be the (unspecified) density function for the prior for the parameters $\gamma_{ig}, \delta_{ig}^2$, and let $L(\mathbf{Z}_{ig} | \gamma_{ig}, \delta_{ig}^2) = \prod_j \varphi(Z_{ijg}, \gamma_{ig}, \delta_{ig}^2)$, where $\phi(Z_{ijg}, \gamma_{ig}, \delta_{ig}^2)$ is the probability density function of an $N(\gamma_{ig}, \delta_{ig}^2)$ random variable evaluated at Z_{ijg}. Using Bayes' theorem, the integral above for can be written as

$$E[\gamma_{ig}] = \frac{1}{C(\mathbf{Z}_{ig})} \int \gamma_{ig} L(\mathbf{Z}_{ig} | \gamma_{ig}, \delta_{ig}^2) \pi(\gamma_{ig}, \delta_{ig}^2) \, d(\gamma_{ig}, \delta_{ig}^2) \qquad (10.4)$$

where

$$C(\mathbf{Z}_{ig}) = \int L(\mathbf{Z}_{ig} | \gamma_{ig}, \delta_{ig}^2) \pi(\gamma_{ig}, \delta_{ig}^2) \, d(\gamma_{ig}, \delta_{ig}^2).$$

We estimate both $C(\mathbf{Z}_{ig})$ and the integral in (10.4) using Monte Carlo integration (Liu 2001; Gilks *et al.* 1996, p. 4) over the empirically estimated pairs $(\widehat{\gamma}_{ig}, \delta_{ig}^2)$, which are considered random draws from $\pi(\gamma_{ig}, \delta_{ig}^2)$. Letting $w_{ig''} = L(\mathbf{Z}_{ig} | \widehat{\gamma}_{ig''}, \widehat{\delta}_{ig''})$ for $g'' = 1, \ldots, G$, $C(\mathbf{Z}_{ig})$ can be estimated by $\widehat{C}(\mathbf{Z}_{ig}) = \frac{1}{n} \sum_{g''} w_{ig''}$ and the integral in (10.4) can be estimated by

$$\gamma_{ig}^* = \widehat{E}(\gamma_{ig}) = \frac{\sum_{g''} w_{ig''} \widehat{\gamma}_{ig''}}{n \widehat{C}(\mathbf{Z}_{ig})}.$$

The same method is used to find the posterior expectation of δ_{ig}^2, and the nonparametric EB batch adjustments $\gamma_{ig}^*, \delta_{ig}^{2*}$ are given by

$$\gamma_{ig}^* = \frac{\sum_{g''} w_{ig''} \widehat{\gamma}_{ig''}}{\sum_{g''} w_{ig''}} \quad \text{and} \quad \delta_{ig}^{2*} = \frac{\sum_{g''} w_{ig''} \widehat{\delta}_{ig''}^2}{\sum_{g''} w_{ig''}}.$$

Using $\gamma_{ig}^*, \delta_{ig}^{2*}$ for the batch adjustment estimates, the data are adjusted using the method described in Step 3 in Section 10.3.1.

10.4 Data Examples, Results and Robustness of the Empirical Bayes Method

10.4.1 Microarray Data with Batch Effects

Data set 1 resulted from an oligonucleotide microarray (Affymetrix HG-U133A) experiment on human lung fibroblast cells (IMR90) designed to reveal whether exposing mammalian cells to nitric oxide (NO) stabilizes mRNAs. Control samples and samples exposed to NO for 1 hour were then transcription inhibited for 7.5 hours. Microarray data was collected at baseline (0 hours, just before transcription inhibition) and at the end of the experiment (after 7.5 hours) for both the control and the NO-treated group. It was hypothesized that NO will induce or inhibit the expression of some genes, but would also stabilize the mRNA of many genes, preventing them from being degraded after 7.5 hours. One sample per treatment combination was hybridized, resulting in four arrays. This experiment was repeated at three different times or in three batches (totaling 12 samples). The batches in this data set were identical experiments using the same cell source, and were conducted by the same researcher in the same lab using the same equipment. These data are available for download at http://statistics.byu.edu/johnson/ComBat/data/.

A common first step in a microarray data analysis is to look at a clustering heat map of the data. Figure 10.1 shows a heatmap of data set 1 using a standard hierarchical clustering algorithm produced using the dChip software (Li and Wong 2003). This heatmap exhibits characteristics commonly seen by researchers attempting to combine multiple batches of microarray data. All four samples in the second batch cluster together, indicating that the clustering algorithm recognized the batch-to-batch variation as the most significant source of variation within this data set. Batches 1 and 3 yield a more desirable result, first

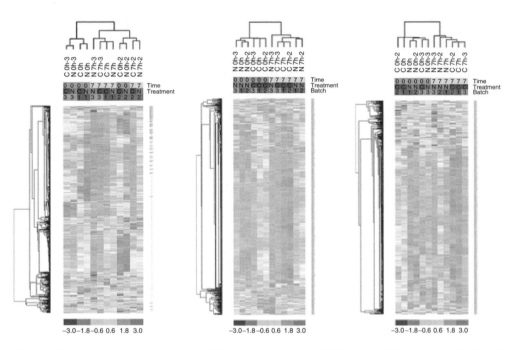

Figure 10.1 A heatmap clustering of data set 1. Genes with large variation across all the 12 samples are filtered and used for clustering, and the gene-wise standardized expression values are used to compute gene and sample correlations and displayed in gray scale. The sample legends on the top are: 0 (0 hour), 7 (7.5 hours), C (Control), N (NO treated). (left) Notice that the samples from the batch 2 cluster together and the baseline (time = 0) samples also cluster by batch 1 and 3. (center) There is no evidence of batch effects after applying an LS adjustment in that samples first separate by Time and then by Treatment, and replicate samples in different batches cluster together. This method is not robust to outliers in small samples sizes. (right) After applying the EB batch adjustments there is little evidence of batch effects. This diagram is similar to the LS heatmap (center), but the EB method does not lose fold changes and is robust to outliers in small sample sizes. (Color version available as Figure 10.1 on website.)

clustering by time and then, in the 7.5 hour cluster, by treatment. However, in the baseline cluster, the samples again cluster by batch. If these data were void of batch effects, one would expect to see clustering only by NO treatment or time. Thus data set 1 is a good example to be considered for adjusting batch effects.

Data set 2 originates from an RNA interference (RNAi) knockout experiment that studied the biological effects of inhibiting the TAL1 oncogene in human cells using an RNAi technique. This experiment was conducted in three separate batches. The first batch contained six RNAi samples and only two controls. After the first batch of the experiment was finished, the researchers determined that the sample sizes, particularly the number of controls, were not sufficient and subsequently added three RNAi samples and four controls. The time delay between the two experiments forced the researchers to hybridize the arrays at different times. Although the arrays were processed at the same facility, strong batch effects were exhibited between the two batches. Therefore the researchers repeated the experiment the third time in one batch with nine RNAi and six control samples. Figure 10.2 contains heatmap clusterings of the samples in all the three batches. The right sample branch mainly consists of samples in batch 3, and the left sample branch mainly consists of samples in batch 1 and 2. These data are available for download at http://statistics.byu.edu/johnson/ComBat/data/.

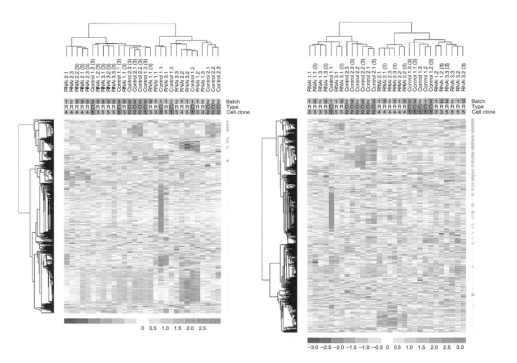

Figure 10.2 A heatmap clustering of data set 2. The sample legends on the top are: C (Control), R (RNAi). (left) There appear to be very prevalent batch effects in these data, particularly for the third batch. (right) After adjustment there is no evidence of batch effects. The samples cluster in small groups based on treatment status (RNAi or control) and cell clone. (Color version available as Figure S10.2 on website.)

The researchers were able to derive some interesting results from the third batch of data set 2 (unpublished data), but the individual sample sizes in the first two batches were considered too small to be useful individually. Using data set 2, we consider whether it is possible to reproduce the results derived solely from the third batch by using only the first two batches of data. Also, we consider the advantage of combining all three batches of data to detect interesting genes.

10.4.2   Results for Data Set 1

The parametric and nonparametric EB adjustments were both applied to data set 1. Both methods produced nearly identical adjustments, as the priors seemed to be moderately reasonable distributions for the priors (Figure 10.3), so the parametric EB adjusted data were used for downstream analysis. Because this was a balanced experiment and the batch sizes were so small (one sample per batch and treatment combination), no covariates were included in the model. Comparing Figure 10.1 (left) to Figure 10.1 (right) provides evidence that the batch effects were adequately adjusted for in these data. Downstream analyses are now appropriate for the combined data without having to worry about batch effects.

Figure 10.4 illustrates the amount of batch parameter shrinkage that occurred for the adjustments for 200 genes from one of the batches from data set 1. The adjustments viewed in this figure are on a standardized scale, so the magnitude of the actual adjustments also depends on the gene-specific expression characteristics (overall mean and pooled variance from standardization) and may vary significantly from gene to gene.

10.4.3   Results for Data Set 2

The parametric forms for the prior estimates (from Step 2 above) were not satisfactory for data set 2, so we applied the nonparametric EB adjustment to these data by first combining the first two batches (call this EB2) and then combining all three batches (EB3). The third batch was used for comparison against the EB2 analysis results because it was an identical experiment to EB2 other than the fact that it was conducted in a single batch. Treatment status (RNAi or control) was included in the EB procedure for these adjustments. For EB3 the samples clustered well into small groups based on treatment status and 14 out of the 15 samples from the third experiment clustered next to their experimental counterpart from the first two studies (see Figure 10.2 (right)).

Differential expression was assessed using Welch's t-test to determine the differential expression of RNAi versus control samples. EB2 produced at list of 86 significant genes at a false discovery (q-value) threshold of 0.05 (Storey and Tibshirani 2003). The third batch alone produced a list of 37 significant genes using the same threshold. Crossing the significant gene lists, we observed 13 genes common in both lists (Fisher's exact p-value $< 10^{-15}$). Furthermore, the dChip software (Li and Wong 2003) was used to find significant Gene Ontology (GO) (http://www.geneontology.org) clusters. To find clusters, we first filtered out the genes that do not satisfy the criterion $c < \text{sd(gene)}/\text{mean(gene)} < 10$, where c is set low enough to allow for about 1000 genes to remain after filtering. This filtering method selects genes with possible differential expression between all conditions,

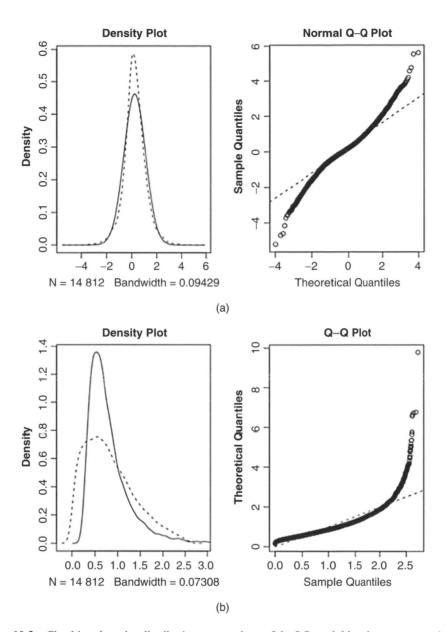

Figure 10.3 Checking the prior distribution assumptions of the LS model batch parameters. (a) The gene-wise estimates of additive batch parameter ($\widehat{\gamma}_{ig}$ of all genes) for data set 1, batch 1. (b) The gene-wise estimates of multiplicative batch parameter ($\widehat{\delta}_{ig}^2$ of all genes) for data set 1, batch 1. Each density plot contains a kernel density estimate of the empirical values (dotted line) and the EB-based prior distribution used in the analysis (solid line). Dotted lines on the quantile–quantile plots correspond to (a) the EB-based normal or (b) the inverse gamma distributions.

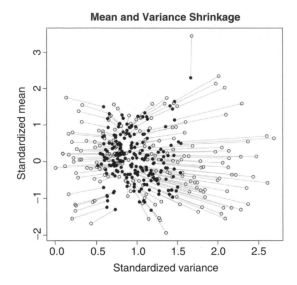

Figure 10.4 Shrinkage plot for the first 200 probes from one of the batches in data set 1. The gene-wise and EB estimates of γ_{ig} and δ_{ig}^2 in Section 10.3.1 are plotted on the y and x axis. Circles are the gene-wise values and the black dots are after applying the EB shrinkage adjustment. (Color version available as Figure S10.4 on website.)

and enrichment of the gene's corresponding GO terms is defined as the overrepresentation of GO terms in the list as compared to the number of occurrences in the data before filtering in all the genes in the array (based on a p-value below 0.05). There were 53 significant GO clusters in EB2 and 36 significant clusters in the third batch. These lists had 32 GO clusters in common.

Welch's t-test was also applied to EB3 to find differentially expressed genes, yielding 1599 genes significant at a q-value cutoff of 0.05, including 34 of the 37 batch 3 genes. Reducing the q-value threshold to 0.01 yielded 488 significant genes (including 32 of the 37 batch 3 genes), and decreasing the threshold further to 0.001 yielded 161 significant genes (including 28 of the 37). The GO cluster analysis for EB3 yielded 85 significant clusters, including 33 of the 36 batch 3 clusters. These results indicated a clear increase in power to detect biological differences and the benefit of combining batches was clearly evident.

10.4.4   Robustness of the Empirical Bayes Method

The robustness of the EB method results from the shrinkage of the LS batch estimates by pooling information across genes. If the observed expression values for a given gene are highly variable across samples in one batch, the EB batch adjustment is more like prior and less like the gene's gene-wise estimates, and becomes more robust to outlying observations. This phenomenon is illustrated in Figure 10.5.

An array with n genes can be thought of as a point in n-dimensional space, and batch adjustments are moving one cluster of points (one batch) to match another using scale

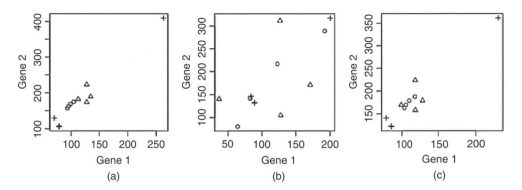

Figure 10.5 Plots illustrating the robustness of the EB adjustments compared to the LS adjustments. The symbols signify batch membership. Data in (a) are unadjusted expression values for two genes from data set 1. Data in (b) are the LS adjusted data. The same symbol corresponds to the samples in the same batch. Notice that the locations and scales of the batches in (b) have been over-adjusted because of outliers in the unadjusted data, and that these outliers disappear in the LS data. Data in (c) are the EB adjusted data. The outliers remain in these data and the batch with outliers is barely adjusted. The batches without outliers are adjusted correctly in the EB data.

and location shifts. The DWD, SVD and LS methods do not work well for small samples because they are selecting multi-dimensional adjustments using only a few points (samples) in multi-dimensional space. In order to visually inspect the effect of batch adjustments, one can consider these data by plotting them on fewer dimensions and still get a reasonable idea of the effect of the batch adjustments. Figure 10.5 contains two-dimensional plots (two genes) from data set 1, selected because both genes contain outlying observations for the same sample. By empirical observation, it seems that there are only small batch effects in the data in these two dimensions (Figure 10.5(a)). Additionally, it seems that the outliers are not caused by batch effects, but are truly an outlying observation in both directions. The outliers highly affect the outcome of the LS estimates, especially the variance (Figure 10.5(b)) as it appears that the variances of the batches without outliers are over-inflated from the adjustment. Without a variance adjustment, there still would be an over-adjustment for the mean shift. However, the EB adjustment (Figure 10.5(c)) works very well in these dimensions. Examples without outlying observations typically show only slight difference between the LS and the EB methods, and in larger samples the LS methods are more robust and the EB and LS methods are usually very similar.

10.4.5 Software Implementation

The EB batch effect adjustment method described here is implemented in the R software package (http://www.r-project.org) and is freely available for download at http://statistics. byu.edu/johnson/ComBat/. Computing times for the batch effect adjustments on data set 1 (3 batches, 12 arrays, 22 283 probes/array) were less than 1 minute for the parametric method and less than 8 minutes for the nonparametric approach on a standard laptop

(Windows XP on a 2.13 GHz Intel Pentium M processor). For data set 2 (3 batches, 30 arrays, 54 675 probes/array) the parametric adjustment took less than 3 minutes, and the nonparametric method took just under an hour to complete.

10.5   Discussion

Batch effects are a very common problem faced by researchers in the area of microarray studies, particularly when combining multiple batches of data from different experiments or if an experiment cannot be conducted all at once. We have reviewed and discussed the advantages and disadvantages of the existing batch effect adjustments. Notably, none of these methods are appropriate when batch sizes are small (less than 10), which is often the case. In order to account for this situation, we have presented a very flexible empirical Bayes framework for adjusting for additive, multiplicative and exponential (when data have been log transformed) batch effects.

We have shown that the EB adjustments allow for the combination of multiple data sets and are robust to small sample sizes. We illustrated the usefulness of our EB adjustments by combining two example data sets containing multiple batches with small batch size, and obtained consistent results from downstream analyses (clusterings and analysis as compared to similar single-batch data) while robustly dealing with outliers in these data.

The aim of the standardization procedure presented in Section 10.3.1 is to reduce gene-to-gene variation in the data, because genes in the array are expected to have different expression profiles or distributions. However, we do expect phenomena that cause batch effects to affect many genes in similar ways. To more clearly extract the common batch biases from the data, the standardization procedure standardizes all genes to have similar overall mean and variance. On this scale, batch effect estimators can be compared and pooled across genes to create robust estimators for batch effects. Without standardization, the gene-specific variation increases the noise in the data and inflates the prior variance, decreasing the amount of shrinkage that occurs. Therefore standardization is crucial for EB shrinkage methods. However this feature is not present in many EB methods for Affymetrix arrays.

In a balanced factorial ANOVA experiment the LS model covariates (β_g in equation (10.1)) are always orthogonal. As a result, the LS additive batch effect estimates are the same whether or not covariates are included in the model, which subsequently implies that the LS adjusted data are the same whether or not covariates are included in the model. Additionally, the standardization procedure has no effect on these LS adjustments, and is therefore unnecessary. Note that this does not apply when numerical covariates are included or when sample sizes or conditions are unbalanced across batches. In contrast, the EB adjustments are sensitive to the inclusion of covariates in the LS model because the residual standard error from the LS model is an important factor in the shrinkage of the LS estimates toward the empirical prior. The objective is to find the best possible estimate for the residual error from the LS model to best estimate the EB batch effect parameters. Therefore if sample sizes are large enough, it is recommended to model all available covariates expected to be significant.

Finally, note that the EB adjustments are dependent on several factors: the standardized batch mean, the empirical prior distribution, the (residual) variance estimate, and the sample size within the batch. Sample size appears to be a very influential factor in this

EB method. As sample sizes increase, the EB adjustment converges to the LS batch effect parameter estimate and diverges from the empirical prior (see equation (10.3)). In data sets were the sample sizes are relatively moderate (15–25 samples per batch) there is usually not much difference between the EB and the LS adjustments. However, since the batch sizes of the example data sets were small, some of the adjustments (particularly those with outlying observations) were strongly influenced by the prior, making them more robust. As shown in the examples above, we conclude that the EB method is advantageous for small sample size because it is less susceptible to outliers in the data.

11

Identical Reference Samples and Empirical Bayes Method for Cross-Batch Gene Expression Analysis

Wynn L Walker and Frank R Sharp

Abstract

Nonbiological experimental error commonly occurs in microarray data collected in different batches. It is often impossible to compare groups of samples from independent experiments because batch effects confound true gene expression differences. Existing methods for adjusting for batch effects require that samples from all biological groups are represented in every batch. In this chapter we review an experimental design along with a generalized empirical Bayes approach which adjusts for cross-experimental batch effects and therefore allows direct comparisons of gene expression values to be made between biological groups drawn from independent experiments. The necessary feature of this experimental design that enables such comparisons to be made is that identical replicate reference samples are included in each batch in every experiment. We discuss the clinical applications as well as the advantages of our approach in relation to meta-analysis approaches and other strategies for batch adjustment.

11.1 Introduction

Nonbiological experimental variation in gene expression values is commonly observed in microarray data processed in different batches. A batch is defined as a set of microarrays processed together within a single experiment. For example, a batch may be a set of microarrays processed on one specific day by a certain operator. In this chapter we define

Batch Effects and Noise in Microarray Experiments: Sources and Solutions edited by A. Scherer
© 2009 John Wiley & Sons, Ltd

an experiment as an individual study conducted at one site during a limited time frame. An experiment often has many samples processed in multiple batches as multiple operators may process different samples. Batch effects are caused by many factors such as the methods for RNA isolation, amplification and target labeling, and array processing and scanning. Several methods have been proposed that can adjust for batch effects provided a large number of samples (at least 25) are included in each batch (Alter *et al.* 2000; Benito *et al.* 2004). More recently, an empirical Bayes method has been described (Johnson *et al.* 2007; see also Chapter 10 in the present volume) that adjusts for batch effects even when the number of samples in each batch is small (less than 10).

The above methods can adjust for batch effects provided that samples from each biological group to be compared are represented in every batch. The left-hand panel of Figure 11.1 shows such an experimental design. In this example experiment, there are two biological groups. It is possible to correct for batch effects in this example because all four batches contain samples from each of the two biological groups (disease and control). In contrast, the experiment in the right-hand panel of Figure 11.1 is an example where it is not possible to distinguish differences in gene expression that are due to batch effects from those that are due to the underlying biology. The reason is that for two of the batches there are samples from only one biological group. Batch 1 is composed entirely of samples from the disease biological group while batch 2 is composed entirely of samples from the control biological group. Therefore, in this example it would not be possible to tease apart the component of the difference in gene expression between batch 1 and batch 2 that is due to biology from that which would be due to the samples being in two separate batches.

Confounding batch effects are even more of a problem when comparing array data from experiments conducted in different laboratories. As a consequence, instead of being able to reuse existing array data processed for control samples researchers need to hybridize the same set of control samples to a new set of arrays and regenerate new data for control samples for each experiment. If it were possible to correct for cross-experimental batch

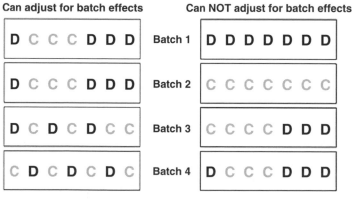

D represents one **disease** sample, C represents one control sample

Figure 11.1   Batch processing of microarray samples from different biological groups. Examples of experimental designs that can (left) and cannot (right) be corrected for batch effects.

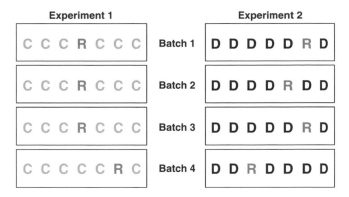

Experiment 1		Experiment 2
C C C R C C C	Batch 1	D D D D D R D
C C C R C C C	Batch 2	D D D D R D D
C C C R C C C	Batch 3	D D D D D R D
C C C C C R C	Batch 4	D D R D D D D

D represents one **disease** sample, C one control sample, and R one reference sample

Figure 11.2 Experimental design with reference samples. This enables the direct comparison of different biological groups drawn from independent experiments that would otherwise be incomparable.

effects, however, data for a single set of patient control samples could be recycled over multiple independent experiments, saving significant resources.

In this chapter we outline an experimental design that enables one to compare data from different biological groups drawn from independent experiments. The feature of our experimental design that allows such comparisons to be made is that *identical* replicate reference samples of pooled RNA are processed in each batch in every experiment. These reference samples are comprised of pooled RNA from multiple individuals drawn from the same tissue – in our study, this tissue is blood. Only a single reference sample is required to be included in each batch. However, in some circumstances, it may be beneficial to include more than one reference sample in each batch. Our experimental design is illustrated in Figure 11.2. In the next section we will review a method for estimating batch effects that is based on the empirical Bayes approach of Johnson *et al.* (2007). We focus on single-channel microarrays, though our methods could easily be extended to two-channel arrays.

11.2 Methodology

11.2.1 Data Description

Gene expression was measured using Affymetrix U133plus2 microarrays on 98 human blood samples. These arrays were processed in two separate experiments. In one experiment, 66 arrays were processed at the University of California at Davis (UC Davis) in six different batches. Fifty-one out of these 66 arrays probed independent samples from teenagers with Duchenne muscular dystrophy ('patients'). The other 15 arrays processed at UC Davis probed the identical reference sample which consists of pooled RNA isolated from the blood of four adults. Five of the batches processed at UC Davis included one reference array, and the sixth batch is composed entirely of reference arrays. In the separate experiment conducted at the University of Cincinnati the control samples were processed. These samples are blood samples drawn from healthy teenagers. A total of 32 arrays were processed in four separate batches, with each batch including between seven and nine arrays. Twenty-eight out of the 32 arrays probed the samples from healthy teenagers ('controls') and the remaining four arrays (one per batch) probed the same pooled reference sample used at UC Davis. Data can be obtained from the www.the-batch-effect-book.org/supplement/data.

Probe level data were summarized into a single expression value for each gene on each array using the GCRMA software in GeneSpring 7 (Agilent Technologies, http://www.chem.agilent.com). Pre-processing involved nonlinear background reduction utilizing probe DNA sequences, quantile normalization, and summarization by median polishing (Bolstad *et al.* 2003; Katz *et al.* 2006). Using the GeneSpring software, we performed unpaired two-sample *t* tests with different mean expression between disease and control groups. The two sample groups were compared for the unadjusted gene expression data as well as the empirical Bayes batch adjusted data. The Benjamini–Hochberg false discovery rate (FDR) method was used to correct the *p*-values for multiple comparisons. We considered an FDR-corrected *p*-value of 0.05 as significant. We further filtered each of the above lists, removing all genes with an average fold change less than 2.0. Differentially expressed genes between batches within each experiment were identified using one-way analysis of variance (ANOVA) within GeneSpring with an FDR-corrected *p*-value threshold of 0.05.

11.2.2   Empirical Bayes Method for Batch Adjustment

Our goal is to identify genes that are differentially expressed in the blood of patients with muscular dystrophy relative to the blood of controls. The mathematical model used for gene expression in the empirical Bayes method (Johnson *et al.* 2007) is

$$Y_{ijg} = \alpha_g + \mathbf{X}\boldsymbol{\beta}_g + \gamma_{ig} + \delta_{ig}\varepsilon_{ijg} \tag{11.1}$$

where Y_{ijg} is the expression value for gene g for sample j from batch i. α_g is the overall average gene expression, $\mathbf{X}$ is the design matrix, and $\boldsymbol{\beta}_g$ is the vector of regression coefficients corresponding to X. γ_{ig} measures the additive batch effect of batch i for gene g which represents the location (mean) of the adjustment for batch i and is assumed to follow a normal distribution. δ_{ig} measures the multiplicative batch effect and is assumed to follow an inverse gamma distribution. The error terms, ε_{ijg}, are assumed to follow a normal distribution with expected value zero and variance σ_g^2.

We employ three versions of the above model to derive three different sets of empirical Bayes batch-adjusted data. The first set of data is adjusted for both the batch effects that occur between batches within each experiment as well as those that occur between batches of the two different experiments (model 1). The second data set is adjusted for only the batch effects that occur between batches from different experiments, and batch effects that occur within either experiment are not considered (model 2). The third data set is adjusted only for batch effects that occur within either experiment, while batch effects that occur between experiments are not considered (model 3). We consider these three separate models in order to assess and compare the relative significance of the two different types of batch effects.

In equation (11.1) for model 1, gene g has parameters γ_{ig} and δ_{ig} for each of the ten distinct batches (six processed at UC Davis and four at the University of Cincinnati). There are a total of three regression coefficients in the vector of parameters for the biological group covariates, β_g, one that specifies whether a sample is the reference, and two that specify disease state (healthy, Duchenne, or non-Duchenne muscular dystrophy).

Model 2 is the same as above except that there are only two values for γ_{ig} and δ_{ig} per gene, one for each of the two experiments. In this model, because within-experiment batch effects are not modeled, there are only two batches; that is, for each experiment all samples are considered to be in the same batch.

For model 3, where the data are not adjusted for between-experiment batch effects, the reference sample is not used and so there are only two coefficients in the vector β_g. The data sets for the two experiments are adjusted for batch effects that occur within the experiment only and therefore are adjusted separately.

The model parameters are estimated as described by Johnson *et al.* (2007). We performed the above calculations using the COMBAT software developed by Johnson *et al.* and written in R (http://statistics.byu.edu/johnson/ComBat). This software features diagnostic plots that allow one to check the validity of the model assumptions.

Differentially expressed genes are identified as previously described (Walker *et al.* 2008).

11.2.3 Naïve t-Test Batch Adjustment

We compare the proposed empirical Bayes approach to a simple method that filters out the identified differentially expressed genes which also have significant cross-experimental batch effects based on a t-test comparing the 15 reference arrays from UC Davis to the four reference arrays processed at the University of Cincinnati. This t-test filter was applied to each gene that was identified as differentially expressed between the diseased and healthy samples in the GCRMA-summarized unadjusted gene expression data set. Each gene that was found to be significantly different between experiments (unadjusted p-value ≤ 0.2) was removed from the gene list because of the ambiguity of whether its differential expression was due to the underlying biology or to the between-experiment batch effects. The loose p-value threshold of 0.2 for removing genes was chosen because of the limited accuracy of the t-test (only four reference arrays were being compared to the other set of 15 reference arrays).

11.3 Application: Expression Profiling of Blood from Muscular Dystrophy Patients

11.3.1 Removal of Cross-Experimental Batch Effects

In our study (Walker *et al.* 2008), we demonstrated that the empirical Bayes method removed batch effects that occur between the two experiments conducted at UC Davis and Cincinnati. We illustrated the successful removal of batch effects by comparing the two hierarchical cluster dendrograms generated from the two different sets of differentially expressed genes – those identified using the unadjusted gene expression values and those identified using the batch-adjusted gene expression values. Each dendrogram was generated by clustering the differentially expressed gene expression values over the entire set of 98 samples. In the dendrograms corresponding to the unadjusted data set's differentially expressed genes, the UC Davis reference arrays and the Cincinnati reference arrays formed distinct clusters. This is indicative of there being a substantial amount of gene expression due to cross-experimental nonbiological artifacts. Identical reference samples otherwise should have similar expression values and so should instead be randomly intermixed amongst each other. Upon adjusting the data for both within and between experiment batch effects (model 1), the reference arrays were mixed in the corresponding dendrogram. This suggests that the cross-experimental batch effects have been removed.

The difference in the numbers of differentially expressed genes found when comparing the sets of reference samples for both the unadjusted data set and the batch-adjusted data set also points to the successful removal of cross-experimental effects. While 85 genes were differentially expressed between the two groups of reference samples for the unadjusted data set, there were no genes found to be differentially expressed between the two sets of reference samples after cross-experiment batch adjustment.

11.3.2 Removal of Within-Experimental Batch Effects

Table 11.1 shows the number of genes within a single experiment for which the average gene expression in at least one batch is significantly different from the other batches (based on a one-way ANOVA with FDR followed by a fold-change filter of 2.0). For example, in the unadjusted data set, 527 out of 54 675 genes were differentially expressed in at least one of the five batches processed at UC Davis, while 48 of the 54 675 genes were differentially expressed in at least one of the four batches processed at Cincinnati. Interestingly, only three genes are in common between the two sets. Table 11.1 shows that there are comparable numbers of genes found to be differentially expressed in at least one batch after adjusting only for between-experiment batch effects (model 2). This indicates that there are a sizable number of genes whose expression values are altered due to the batch effects that occur within either experiment.

The corresponding numbers of differentially expressed genes are shown in Table 11.1 for the data sets adjusted for both types of batch effects (model 1) and for within-experiment batch effects only (model 3). For the data set adjusted only for

Table 11.1 Numbers of genes found to be differentially expressed in at least one batch within either experiment based on ANOVA. Numbers are out of 54 675 probe sets unless otherwise specified. EB = empirical Bayes.

Adjustment method	Cincinnati	UC davis	Common
None	48	527	3
t-test filter	0 (of 273)	42 (of 273)	0
EB: between experiment only	96	545	8
EB: within experiment only	0	4	0
EB: within and between experiment	0	0	0

within-experiment batch effects, there are only four genes that are found to be differentially expressed in at least one batch relative to all other batches processed at UC Davis. These small numbers show that the empirical Bayes method is very effective at removing within-experiment batch effects. When the data was adjusted for both types of batch effects, there were no differentially expressed genes across batches.

11.3.3 Removal of Batch Effects: Empirical Bayes Method versus t-Test Filter

In this section we discuss the advantages that the empirical Bayes method offers by comparing it to a more naïve approach which uses a *t*-test to filter out genes with significant differences in expression between reference arrays in the two experiments. Table 11.2 shows the genes identified as differentially expressed in patients relative to controls for the different methods explored in this paper: method 1, adjustment for both types of batch effects (model 1: 629 genes); method 2, *t*-test filtering (273 genes); and method 3, unadjusted data (527 genes). Note that applying the *t*-test as a filter to the 527 genes identified as differentially expressed between patients and controls in the unadjusted data results in the removal of 254 genes. This indicates that nearly half of the differentially expressed genes are likely to be artifactual. Clearly, the relatively small number common to all three gene lists (239 genes) illustrates the profound effect of correcting for batch effects.

It is also interesting to note that 239 out of 273 of the genes identified by the *t*-test filter were also identified by the empirical Bayes method. This suggests that the empirical Bayes method successfully identifies most of the genes that were identified by the *t*-test filter. The empirical Bayes approach also identifies many more genes because it adjusts the values in the gene expression matrix instead of simply removing genes from a list. This gives it the ability to recover false negatives – genes whose differential expression

Table 11.2 Numbers of differentially expressed genes identified by different methods: (1) Empirical Bayes; (2) *t*-test filtered; and (3) unadjusted data. There are 239 genes common to all three gene lists. Numbers are out of 54 675 probe sets unless otherwise specified.

Adjustment method	Number of genes identified
All genes found by method 1	629
All genes found by method 2	273
All genes found by method 3	527
Genes found only by method 1	272
Genes found only by method 2	0
Genes found only by method 3	136
Genes found by both methods 1 and 2	239
Genes found by both methods 1 and 3	357
Genes found by both methods 2 and 3	273
Genes found by all three methods	239

values would be artificially canceled out by an opposing differential expression due to batch effects. Finally, out of the 273 genes which passed the filter, a significant number, 42 genes, show differential gene expression in at least one of the five batches processed at UC Davis (Table 11.1). This is not surprising because the t-test removes genes with between-experiment batch effects but does not adjust for batch effects *within* an experiment.

11.4   Discussion and Conclusion

11.4.1   Methods for Batch Adjustment Within and Across Experiments

A variety of methods have been developed to correct for nonbiological variation due to batch effects within an experiment and across experiments. Alter *et al.* (2000) used singular value decomposition to adjust for nonbiological variation in yeast cell cycle data. Another group (Nielsen *et al.* 2002) applied a similar method to correct for batch effects in a tumor data set. Alter *et al.* adjusted for nonbiological variation by inferring that the combinations of genes and arrays that contributed the most to the variance correspond to nonbiological artifacts. They normalized the data by filtering out these combinations of genes and arrays and compared expression values for the remaining genes using the remaining arrays. This method succeeded in this case because the nonbiological variation happened to be the greatest source of variation. However, if the amount of biological variation happened to be greater than the nonbiological variation, this approach would have failed to discern the nonbiological variation. This motivated others (Benito *et al.* 2004) to apply a different approach, distance weighted discrimination, to correct for systematic batch effects. For further description of this method and its application, see Chapter 13.

Yet another method is the surrogate variable analysis (SVA), developed by Leek and Storey (2007). SVA was applied to gene expression studies of disease classes, time-course experiments and genetic dissection of expression variation. The concept of SVA is the generation of surrogate variables as covariates for regression analysis on the residual data matrix, after the signal due to the primary source had been removed. This allows for the identification of signatures of experimental heterogeneity as singular vectors which represent this variation and estimate their relation to the primary variable.

Sims *et al.* (2008) chose a batch mean-centering approach for adjustment of batch effects across experiments. In this method, the mean expression level per probe set in a given data set is subtracted from the individual GeneChip expression level on the $\log_2$ scale. The authors employed this method to compile a large data set of 1107 arrays from six published studies and show improved concordance and more accurate predictions.

The Bayesian approach developed by Johnson *et al.* (2007) is based on a location and scale model that allows a different mean and variance to be modeled for each gene and batch (for a review, see Li and Wong 2003). This method pools information across genes in each batch when estimating the model parameters. Therefore, it does not require a large number of samples to be processed in each batch in order to provide robust batch adjustments for each gene. See Chapter 10 for further details and discussion of this method.

11.4.2 Bayesian Approach is Well Suited for Modeling
         Cross-Experimental Batch Effects

The Bayesian approach is particularly advantageous for the analysis of our data set because it had a small number of reference samples – most batches included only one reference sample. Because the pooled variance for each gene is calculated across *all* samples in the parameter estimation procedure, the multiplicative batch effect (i.e. the variance) can still be estimated even when there is only one reference sample per batch. An experimental design would ideally include more than one reference sample per batch. The presence of more replicate samples would provide a more robust estimate of the batch effects. Along these lines, one would be able to estimate batch effects in cases where one of the samples in the batch was of poor quality. In our study we were able to justify the use of a single reference sample based on the results of the ANOVA analysis of the batch-adjusted data (Table 11.1). The bottom row of this table shows that after adjustment of batch effects there are no differentially expressed genes between batches for either of the two experiments. However, if a reference sample in a batch deviated from the other reference samples, one might expect that upon batch adjustment there would be artifactual differences in gene expression values between the genes in the corresponding batch and the remaining batches. This was clearly not the case in our example.

A similar concern is that the variance for many genes might be unusually large relative to the differences in the batch means because only one reference sample is included in each batch. If this were indeed the case, the adjustment method would oftentimes only add noise and reduce sensitivity. For our data set, the observed estimated parameter values for the batch variances were infrequently large relative to the differences in the batch means. Again, even though there is only one reference sample per batch, the variance estimate is likely stabilized by the pooling of samples.

11.4.3 Implications of Cross-Experimental Batch Corrections
         for Clinical Studies

We have shown that with an appropriate experimental design and statistical methods it is possible to adjust for both within- and between-experiment batch effects and there-fore compare gene expression values between biological groups drawn from independent experiments using the same platform. The unique feature of our experimental design that enables us to compare data from separate experiments is that identical replicate reference samples of pooled RNA (derived from the same tissue as the experimental samples) are included in each batch within each experiment. The inclusion of the identical reference samples from the same tissue in every batch in every experiment allows us to adjust for nonbiological variation and hence to isolate the differences in gene expression that are due to the underlying biology from those due to the nonbiological batch effects. Because gene expression values from data sets drawn from many independent experiments could be compared, there is enormous increase in the number of hypotheses that could be tested with available data sets.

Our proposed experimental design also offers a huge benefit for the use of microarrays in clinical studies. Because, with the use of identical reference samples, it is possible to compare gene expression values for disease and control samples drawn from separate experiments, it is possible to recycle expression data for one set of controls. This would not only eliminate the burden and expense of reprocessing control samples each time an experiment is repeated but also eliminate need to recruit a new control group each time an experiment is conducted. If it were possible to recycle gene expression values for a single control gene expression data set, it would become possible to define a single standard control population for a disease study and use the gene expression values derived from this population as a universal reference data set. This would dramatically increase the utility of microarrays in clinical diagnostics.

12

Principal Variance Components Analysis: Estimating Batch Effects in Microarray Gene Expression Data

Jianying Li, Pierre R Bushel, Tzu-Ming Chu, and Russell D Wolfinger

Abstract

Batch effects are present in microarray experiments due to poor experimental design and when data are combined from different studies. To assess the quantity of batch effects, we present a novel hybrid approach known as principal variance components analysis (PVCA). The approach leverages the strengths of two popular data analysis methods: principal components analysis and variance components analysis, and integrates them into a novel algorithm. It can be used as a screening tool to determine the sources of variability, and, using the eigenvalues associated with their corresponding eigenvectors as weights, to quantify the magnitude of each source of variability (including each batch effect) presented as a proportion of total variance. Although PVCA is a generic approach for quantifying the corresponding proportion of variation of each effect, it can be a handy assessment for estimating batch effect before and after batch normalization.

12.1 Introduction

Along with the advance of modern molecular biology, microarray gene expression analysis has become a powerful tool to monitor biological responses on the genome scale (Schena *et al*. 1995). In the past decade, this technology has become a standard procedure to explore basic biological questions as well as to investigate specific human diseases such as cancer. Recent efforts at the Toxicogenomics Research Consortium (led by the National Institute of Environmental Health Sciences (NIEHS) National Center for

Toxicogenomics (NCT)) and the MicroArray Quality Control Consortium (MAQC, led by the Food and Drug Administration) have focused on identifying the sources of variability in gene expression studies and determining the transferability and predictability across array platforms, research laboratories, and tissues to possibly provide a foundation for the use of genomics in future clinical applications (Alter *et al.* 2000; Qin and Kerr 2004; Guo *et al.* 2006; Shi *et al.* 2006).

Many researchers have emphasized the importance and necessity of applying batch correction prior to microarray data analysis (Alter *et al.* 2000; Benito *et al.* 2004; Fan *et al.* 2006; Shabalin *et al.* 2008). Common approaches include mean/median centering or analysis of variance (ANOVA)-like modeling to balance the expression measurement signals across experiments. More sophisticated procedures have also been developed and utilized, including singular value decomposition, distance weighted discrimination, and a more recent method using an empirical Bayes method (Tibshirani *et al.* 2002; Benito *et al.* 2004; Johnson *et al.* 2007). Currently, there is no standard way of assessing the existence and magnitude of batch effects within microarray gene expression data. On the one hand, one can easily overlook such an important issue and directly move on to subsequent data analysis; on the other hand, researchers have sometimes over- or underestimated batch effects and improperly normalized their data. Taking either step without properly estimating the sources and magnitude of batch effects will inevitably result in the potential loss of salient biological information and/or a distorted interpretation of the analytical results. As one can imagine, this can eventually cause an unnecessary waste of time, money and resources and could potentially result in misleading conclusions.

To view sources of variability in gene expression data, a popular approach is to apply principal components analysis (PCA) to the data and then color and/or mark sample scores in two- or three-dimensional space. Such an approach, while valuable, is *ad hoc* and can be very tedious when numerous potential sources of variability are in the data. It also fails to provide a quantitative estimate of how strong a potential source of variability may be compared to other potential sources. Doing so runs the risk of misleading results, potentially leading to errant interpretation of the biological information.

An intuitive proposition is a novel approach called principal variance components analysis (PVCA), which estimates the magnitude of each source of variability and visually compares standardized variance components estimates (see Figure 12.1 for a flowchart). This strategy involves four basic steps: (1) perform PCA to reduce the dimension and retain the majority of the variability in the expression data, say with the first few principal components; (2) fit a mixed model separately to each principal component with all factors of interest as random effects and any nuisance factors as fixed effects; (3) for each factor, average the estimated variance components with their corresponding eigenvalues as weights; (4) standardize the weighted average variance components estimates by dividing by their sum, so that the magnitude of each effect can be represented as a proportion of the total variance.

PVCA has several appealing features. First, the modeling procedure is effectively invariant to the order of the factors. Thus, any potential confounding issue between factors will typically not interfere with the variance estimation, as all factors compete to explain variability. Second, the variance component for each factor will be estimated through restricted maximum likelihood (REML), which is the most efficient and accurate means to estimate variance components (Littell *et al.* 1996), especially when an unbalanced experimental

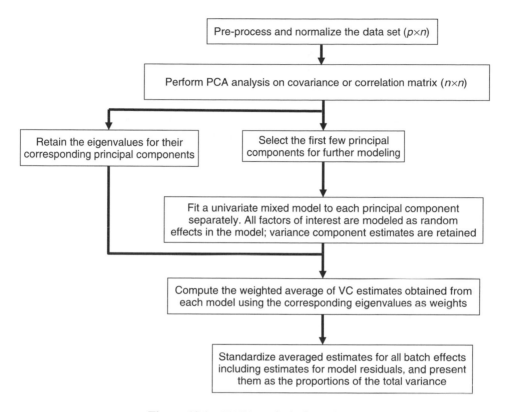

Figure 12.1 PVCA analysis flow chart.

design is involved. Third, variance components analysis (VCA) can also be utilized to estimate the model residual, providing a convenient way to summarize unexplained variance and assess its magnitude against all other known factors in the model.

12.2 Methods

12.2.1 Principal Components Analysis

Data acquired from microarray gene expression technology exhibits an important statistical phenomenon – high dimension (large p) and low sample size (small n). Here we are assuming that there are p probes (genes) fabricated on the array and there are n arrays applied for different biological conditions (chemical, dose level, time duration before the animal was sacrificed, etc.) observed in the experimental design. Many traditional multivariate analysis procedures become useless since the inverse of the covariance matrix of the data is massively singular. One practical solution is to seek some type of dimension reduction procedure. PCA has two major features. First, through algebraic projection, it represents the original data in a new data space with the same order; however, the axes (principal components) in the new data space are orthogonal to each other. Second, the newly formed axes are ordered in a sequentially reduced fashion in terms of their weight; that is, any given principal component captures more variability than the one that immediately follows it. Below is an overview of PCA to illustrate the concept, basis, and actual performance of the approach. Any good linear algebra textbook or multivariate statistics resource will provide more details.

As mentioned above, in a microarray gene expression experiment researchers often deal with a $(p \times n)$ data matrix, where p indicates total number of probes on an array and n represents the number of arrays applied. A random vector $\mathbf{X}' = [X_1, X_2, \ldots, X_n]$ is used, where each array-associated random variable x_i has n observations. The random vector $\mathbf{X}'$ has the variance–covariance matrix $\mathbf{\Sigma}$. Going through all n random vectors, the variance–covariance matrix is expressed as

$$\mathbf{\Sigma} = \mathrm{Cov}(\mathbf{X}) = \begin{bmatrix} \sigma_{11} & \sigma_{12} & \cdots & \sigma_{1n} \\ \sigma_{21} & \sigma_{22} & \cdots & \sigma_{2n} \\ \vdots & \vdots & \ddots & \vdots \\ \sigma_{n1} & \sigma_{n2} & \cdots & \sigma_{nn} \end{bmatrix}. \tag{12.1}$$

The diagonal contains the variance measures for each random variable and the off-diagonal all pairwise covariance measures. In an ideal situation (no systematic error and/or batch effect), this variance–covariance matrix contains the important biological information. Through the application of sophisticated statistical procedures, one can reveal such information by dissecting and manipulating such a matrix.

Starting from our $(n \times n)$ variance–covariance matrix $\mathbf{\Sigma}$, we try to find a list of n scalars (eigenvalues) $\lambda_1, \lambda_2, \ldots, \lambda_n$ satisfying the polynomial equation $|\mathbf{\Sigma C} - \lambda \mathbf{C}| = 0$, where $\mathbf{C}$ denotes an $(n \times n)$ matrix of eigenvectors. These scalars are also sorted in the order such that $\lambda_1 \geq \lambda_2 \geq \cdots \geq \lambda_n \geq 0$. Next, for each eigenvalue λ_i, the corresponding $(n \times 1)$ eigenvector $\mathbf{e}_i$ can be extracted by solving $\mathbf{\Sigma e}_i = \lambda_i \mathbf{e}_i$. The last step is to obtain the principal components by projecting data matrix $\mathbf{X}$ onto the corresponding eigenvector. For each eigenvalue–eigenvector pair $(\lambda_i, \mathbf{e}_i)$, the newly formed ith principal component is given by

$$\mathrm{PC}_i = \mathbf{Y}_i = \mathbf{e}_i' \mathbf{X} = e_{i1} X_1 + e_{i2} X_2 \cdots + e_{in} X_n. \tag{12.2}$$

In theory, eigenvalue λ_i represents the variance associated with the ith principal component, while the sum of all eigenvalues, $\sum \lambda_i$, equals the total variance of the data matrix. Also, through such a procedure to obtain eigenvalues and eigenvector pairs, the principal components are mutually orthogonal, which implies that the covariance between two principal components is zero: $\mathrm{Cov}(\mathbf{Y}_i, \mathbf{Y}_j) = 0$. See Johnson and Wichern (2002) for details. Algebraically, the newly formed principal components are linear combinations of the original random variables $X_1, X_2, \ldots, X_n$ with a constraint that the first principal component carries the highest proportion, $\lambda_1/\sum \lambda_i$, of variability and the next component carries the second highest proportion, $\lambda_2/\sum \lambda_i$, of variability, and so on. Since the principal components are ordered according to their eigenvalues, it is sufficient to select the first x principal components $(x < p)$ containing an amount of variation larger than a predefined percentage threshold (say, 60–90%).

12.2.2 Variance Components Analysis and Mixed Models

In an experimental design where the researcher is mainly interested in effects of the specific factor levels chosen, these factors are defined as fixed effects. For example, in a two-color microarray platform, samples are labeled either with Cy3 (green) or Cy5 (red) fluorescent dye. Due to the chemical structure difference, the labeling efficiency differs between these two dyes. In this case, the dye effect is a fixed effect and we are only interested in the mean effect from either dye. There is a second class of effects, which are considered to be sampled randomly from a population. For example, when different technicians perform a microarray experiment, their different proficiency levels often lead to different results. Here, we often care more about the variability introduced by the technician's general performance, considered within a population of technicians. Since these effects are sampled from a population, the variance of those effects that contribute to the overall variability of the model becomes the key component of interest. Those effects are defined as random effects and can be studied with a different scope. The statistical model to fit an experiment design that includes both fixed and random effects is called a mixed model, and the variance of each random effect is called a variance component. A standard procedure for estimating variance components is described below.

In a general linear model $Y = X\beta + e$, where y denotes the vector of observations, X is the known matrix for each X_{ij}, β is the known fixed-effects parameter vector, and e is the unobserved vector of independent and identically distributed (i.i.d.) Gaussian random errors, with $V_e = \sigma^2 I$. The general form of a mixed linear model is $Y = X\beta + Zu + e$, where Z is the design matrix for random effects, u is the vector of unknown random effect parameters, and e is the unobserved vector of i.i.d. Gaussian random errors. Here, we assume that u and e are normally distributed with

$$E\begin{bmatrix} u \\ e \end{bmatrix} = \begin{bmatrix} 0 \\ 0 \end{bmatrix} \text{ and } \text{Var} \begin{bmatrix} u \\ e \end{bmatrix} = \begin{bmatrix} G & 0 \\ 0 & R \end{bmatrix}.$$

Given that the variance of Y is $V = ZGZ' + R$, V can be modeled by setting up the random effects design matrix Z and by specifying the variance–covariance structure for G and R. In the usual variance components models, G is a diagonal matrix with variance components on the diagonal, each replicated along the diagonal corresponding to the design matrix Z. R is simply the residual variance component times the $n \times n$ identity matrix. Thus, the goal becomes to find a reasonable estimate of G and R. The REML method is the standard procedure, and SAS PROC MIXED is a standard tool for performing REML estimation of variance components (see Littell *et al.* 1996; or http://support.sas.com/documentation/).

12.2.3 Principal Variance Components Analysis

With the fundamentals of PCA and VCA behind us, we introduce a novel hybrid approach integrating the two aforementioned procedures to estimate the possible batch effects embedded in the experimental data. Through PVCA the proportion of the variance components explained by each effect is standardized so that each effect can be displayed as a proportion of the total variance in the data. As illustrated in the flowchart in Figure 12.1, the first step is to obtain the variance–covariance matrix of the $(p \times n)$ microarray expression data matrix. The resulting variance–covariance matrix is a square matrix $(n \times n)$. An alternative option for the first step here is to construct the correlation matrix instead of the variance–covariance matrix. A correlation matrix is a variance–covariance matrix on standardized X_is by subtracting the within-array mean and dividing by the within-array standard deviation. The resulting matrix has 1s on the diagonal and pairwise correlation coefficients in the off-diagonal positions. This then allows, for example, Pearson correlation coefficients to be computed:

$$\rho_{x_i, x_j} = \frac{\mathrm{Cov}(X_i, X_j)}{\mathrm{Var}(X_i)\mathrm{Var}(X_j)}.$$

Secondly, PCA is applied to the correlation or covariance matrix. This procedure is performed in a sequential manner according to the descending eigenvalue order. To determine the number of principal components that are retained for the following procedure, we recommend two criteria: (1) setting the maximum number of principal components to be used in the following mixed model at 10; and (2) requiring the total variance explained by the selected principal components to be above a predetermined threshold. When either criterion is met, the principal components are retained and used in the following steps. In practice, one can also examine scree plots or perform tests based on the Tracy–Widom distribution.

Once the number of principal components is determined, a variance component model is fitted separately to each of these principal components, considering a particular principal components score vector as a univariate response. All effects as well as interactions of interest are considered as random effects in the mixed model. One can optionally include as fixed effects any factors that are not to be considered in the variance partitioning but which are used to adjust the data. All the variance components estimated for each factor are then averaged using the corresponding eigenvalues as weights. A weighted average estimate of the model residual variance is also computed. Finally, for all sources of variability (including batch effect), the averaged estimates are standardized by dividing by their sum to yield a proportion-of-total variability estimate for each factor, potential interaction terms, and the residual variance. These proportions can be conveniently displayed as bar or pie charts.

12.3   Experimental Data

This section describes three publicly available microarray data sets that we will use as examples. PVCA is applied to these data sets to exemplify the usability and the power of this approach to estimate and quantify effects embedded in the data.

12.3.1   A Transcription Inhibition Study

Recently a new batch correction approach (Johnson et al. 2007) based on empirical Bayes methods has exhibited robust properties for batch correction even when the sample size is fairly small. The microarray data used by Johnson et al. involves two factors: exposure to nitric oxide (NO) and transcription inhibition. The experiment was performed in three batches on the Affymetrix oligonucleotide microarray (HG-U133A) platform. Human lung fibroblast cells (IMR90) were exposed to NO for 1 hour and then subjected to a transcription inhibition procedure. The motivating hypothesis is that pre-exposure to NO helps to stabilize the mRNAs. Cells were harvested directly after exposure both before transcription inhibition (0 hours) and at the end of transcription inhibition (7.5 hours). The experiment was repeated in three batches and a total of 12 arrays were used. For further details, see Johnson et al. (2007). Data and PVCA R script are available within the file Rscript.zip on http://www.niehs.nih.gov/research/resources/software/pvca; or from the book companion website www.the-batch-effect-book.org.

12.3.2 A Lung Cancer Toxicity Study

In the MAQC phase II project, 36 working teams from the private sector, academia, and government research institutes worked together to investigate and validate the potential usage of microarray data in clinical diagnosis. Among several microarray experiments, two from Thomas et al. (2007a, 2007b) were used in constructing prediction models, while the third set from the same group was used to validate and assess the specificity and sensitivity of the models. Here, we applied PVCA on these two publicly available data sets. Of the 13 chemicals used in the study, seven were positive for an increased incidence of primary alveolar/bronchiolar adenomas or carcinoma and six were negative. The experiment was performed across a two-year time frame (2005–2006), with two positive chemicals and two negative chemicals tested in one year while the rest were tested in the other. Following 13 weeks of exposure, the mice were euthanized with a lethal intraperitoneal dose of sodium pentobarbital and total RNA was extracted from lung samples. The microarray experiment was performed with a standard protocol (for details, see Thomas et al. 2007a, 2007b); labeled cRNA was fragmented and hybridized to Affymetrix Mouse Genome 430 2.0 arrays. A total of 70 arrays were involved in this experiment (data available at http://www.ncbi.nlm.nih.gov/geo; project codes. GSE6116, GSE5127 and GSE5128).

12.3.3 A Hepato-toxicant Toxicity Study

A comprehensive toxicogenomics study was carried out at the NIEHS-NCT (Lobenhofer et al. 2008) to study acute liver injury upon exposure to a series of eight hepatotoxicants at different dosing levels (Table 12.1). The seven compounds selected were acute hepato-toxicants, and one compound was a nontoxic analog at low doses but toxic at high doses. Male Fisher 344/N rats at approximately 9–12 weeks old were administered with one of the eight chemical compounds; each chemical compound tested was given at three or four different dose levels; rats were sacrificed at 6, 24, or 48 hours after exposure. RNA samples were extracted from the rat liver lobe and further amplified and labeled with fluorescent dyes according to vendor recommended protocol (Agilent Technologies, Palo Alto, CA, USA). Equal amounts of Cy3- or Cy5-labeled cRNA were hybridized to the Agilent Rat Oligo Microarray (Agilent Technologies; for details, see Lobenhofer et al. 2008). In total 318 rats were involved in this study. In order to deal with dye effects, dye swaps were used as part of the experimental design. Therefore, a total of 636 arrays were used in this experiment (data available at http://www.ncbi.nlm.nih.gov/geo; project code: GSE15785).

Table 12.1 Hepatotoxicants used in NCT/NIEHS liver toxicity study. Note that 1,4-dichlorobenzene is a nontoxic analog of 1,2-dichlorobenzene at low doses. All chemicals were given to rats via oral gavage or intraperitoneal injection, and animals were sacrificed at 6, 24, or 48 hours after exposure. Four rats were designated for each compound except diquat at a given dose/time experimental condition. In the event of a rat death, the study was continued with three rats for that group.

Compound	Dosing level (mg/kg)	Time (hour)	Number of rats
Bromobenzene	25/75/250	6/24/48	4
1,2-Dichlorobenzene	15/150/1500	6/24/48	3/4
1,4-Dichlorobenzene	15/150/1500	6/24/48	4
N-nitrosomorpholine	10/50/300	6/24/48	4
Diquat	5/10/20/25	6/24/48	6
Monocrotaline	10/50/300	6/24/48	3/4
Thioacetamide	15/50/150	6/24/48	4
Galactosamine	25/100/400	6/24/48	4

12.4   Application of the PVCA Procedure to the Three Example Data Sets

PVCA was performed using JMP Genomics software (www.jmp.com/genomics). All figures were generated in JMP Genomics. With these three data sets, we are able to show appropriate results and provide a proof-of-concept for applying PVCA procedure prior to detailed data analysis.

12.4.1   PVCA Provides Detailed Estimates of Batch Effects

To test the hypothesis that exposure to NO helps to stabilize mRNA, both sample cells exposed to NO and those not exposed to NO were subjected to a transcription inhibition procedure (time 7.5 hours). For comparison, another set of cells with the same NO exposure pre-condition were analyzed without transcription inhibition (time 0 hours). Microarray data were kindly provided by the corresponding author on the research paper (Johnson *et al.* 2007) and were further $\log_2$ transformed. Following the procedure described in the paper, 752 variant genes were obtained from the ANOVA model. Hierarchical cluster analysis (Figure 12.2) performed on this data set with 752 genes clearly showed three major groups (highest dendrograms), these three groups corresponding to the batches in the experiment. Surprisingly, the main factors that the researchers were interested in (NO exposure and transcription inhibition) within a batch did not show a consistent pattern. For batch 1, two samples exposed to NO clustered more closely; whereas for batches 2 and 3, two samples undergoing transcription inhibition appeared to cluster more closely. Although the hierarchical clustering result is able to show the cluster pattern, it does not provide information about how significant the batch effect is nor a quantitative estimate of it. Based on the PCA on the subset of 752 genes, the first three components explain 59.6% of variability in data, and they were used in the mixed model as univariate

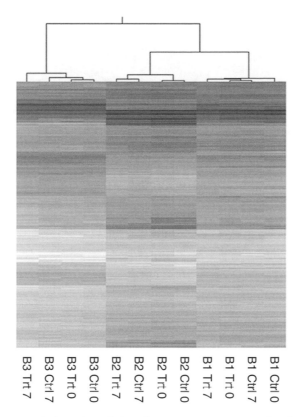

Figure 12.2 Heatmap and clustering of data from a recent paper on batch correction (Johnson *et al.* 2007). The treatment was exposure to nitric oxide (NO) for 1 hour (control cells were not exposed to NO, ctrl); transcription inhibition was done for 7 hours (0 for no transcription inhibition); the experiment was done in three separate batches (B1, B2, B3). Rows are the 752 most variant genes, and columns are experimental conditions.

responses to estimate the variance components (PVCA). In the mixed model, batch, NO exposure (treatment), and transcription inhibition (time), as well as pairwise and three-way interactions, were all modeled as the random effects. A pie chart shows the standardized partition of total variance (Figure 12.3), where two major sources of variability contributed about 87.72% variability in the data (batch effect captures 49.89% of variability and transcription inhibition captures 37.83%). After the application of the empirical Bayes (ComBat) batch correction procedure proposed by the author, only subtle batch effects were detected by PVCA, where the main experiment factor can be modeled to address the questions of biological interest (data not shown).

12.4.2 Visualizing the Sources of Batch Effects

In another example, we applied PVCA analysis to the lung cancer microarray data, which had been pre-processed by data transformation and normalization. From the scree plot

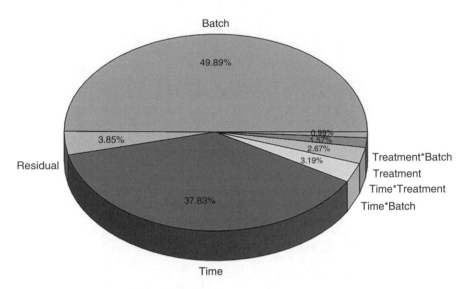

Figure 12.3 A pie chart produced from PVCA analysis on data in Johnson *et al.* (2007). All sources of variability were determined from the modeling and standardized to the total. The first three components (which explained about 59.6% of the variability in the data) were used as the univariate responses variables in the mixed model to estimate the quantity of variance components. The relative percentage associated with each factor is shown.

(Figure 12.4(a)), we observe an elbow shape at the third principal component, which implies that the principal components after the third do not contribute much to the total variability of the data. Based on what is shown in the scree plot, variance component estimation was further obtained by fitting the mixed model on the first three principal components separately and summarizing accordingly. Finally, the standardized variance components estimates are displayed in a bar chart (Figure 12.4(b)). As shown in the bar chart, the batch effect introduced by combining two studies done in two different years comprises the largest proportion (50.9%) of total variability, which was more than double the amount of variability contributed by chemical compounds (20.8%). Such a bar chart provides an intuitive visualization of the major sources of variability in the experimental data. A three-dimensional principal component plot was also produced (Figure 12.4(c)). In this plot, experimental rats were clustered according to the time when the microarray experiment was performed, suggesting a potential batch effect. The first three principal components together contributed 49.5% of the total variability.

12.4.3 Selecting the Principal Components in the Modeling

For the study conducted at NIEHS, all eight chemicals were administered at three dose levels except diquat, where four dose levels were used. Therefore, we merged the middle two dose levels as the medium dose level to balance the experiment. Microarray data were pre-processed by $\log_2$ transformation and an ANOVA-based normalization prior to PVCA analysis. From the scree plot (not shown), it was difficult to determine a plateau pattern or

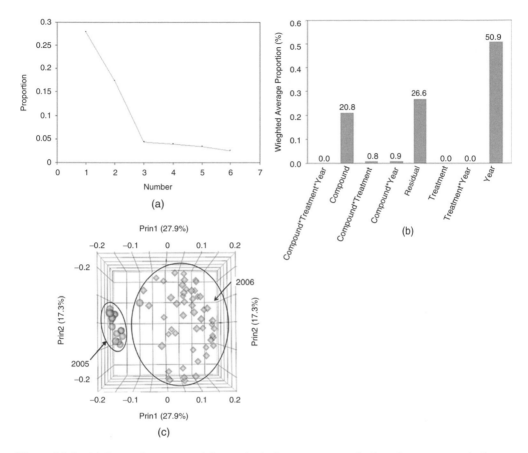

(a)

(b)

(c)

Figure 12.4   (a) Scree plot generated from principal component analysis on lung cancer microarray data (Thomas *et al.* (2007a, 2007b)). It shows an elbow-shaped drop in the contribution after the third principal component. The X-axis is the principal component number, and the y-axis is the proportion of variability explained by the corresponding principal component. (b) PVCA analysis was applied to a lung cancer microarray experiment done at two different times (2005 and 2006). Variance components were estimated and summarized on the first three principal components obtained from the PCA procedure. A total of 13 chemical compounds was used in the combined study; 'treatment' refers to the chemical category (liver carcinogen, liver no-carcinogen, and vehicle control). (c) The plot generated from the PVCA analysis shows the clustering of experimental sample (rat) from two different batches of the study (2005 and 2006).

elbow shape. Therefore, we chose the first 10 principal components for PVCA. Together, these capture 50.1% of the total variability in the data (Table 12.2). According to the PVCA algorithm described in the methods, all factors as well as their interaction terms were treated as random effects in the mixed model for variance component estimation. In Figure 12.5 the residual (unexplained) variance is seen to be the largest component, in agreement with the need for a large number of principal components. Besides the residual, the main factors and interactions are displayed as major contributors to the overall total variability of the data. They are chemical (18.7%), dye (13.4%), time and chemical

Table 12.2 PCA was performed on NIEHS microarray experiment data (Lobenhofer *et al.* 2008). Shown here are the eigenvalues associated with their corresponding eigenvectors from where the principal components were constructed. The proportion column is the proportion of the variability captured by each principal component, and the cumulative amount of variability captured by principal components is also provided.

Principal component	Eigenvalue	Proportion	Cumulative
PC-1	128.386	0.101	0.101
PC-2	110.054	0.087	0.187
PC-3	94.503	0.074	0.262
PC-4	82.746	0.065	0.327
PC-5	54.948	0.043	0.370
PC-6	51.430	0.040	0.410
PC-7	37.358	0.029	0.440
PC-8	32.088	0.025	0.465
PC-9	25.390	0.020	0.485
PC-10	20.777	0.016	0.501

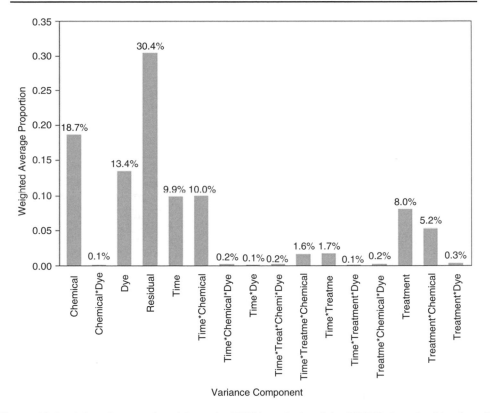

Figure 12.5 A bar chart produced from the PVCA analysis of the NIEHS data. In this plot, all single effects and their possible interactions (two-way, three-way, and four-way) were estimated for their contribution to the overall variability in the data. The sum of all variance components was set to 1.

interaction (10.0%), time (9.9%), treatment (8.0%), treatment and chemical interaction (5.2%), etc. However, there was no obvious batch effect detected. Clearly, those factors and interaction terms were the targets and of major interest when designing the experiment. Our PVCA algorithm highlights them from a different perspective and quantitatively compares them to the overall unexplained sources of variability.

12.5 Discussion

Since the biological samples or specimens in an experimental study are occasionally a limited resource, using microarray analysis to find informative genes presents a statistical challenge – that of high dimension (large p) and low sample size (small n). As a result, researchers may suffer from the 'curse of dimensionality' in microarray data analysis. PCA offers a natural solution, in that it searches for orthogonal axes (principal components) of the original data. One important constraint is that the newly formed principal components must carry partial variability in the original data in a sequentially decreasing order. Oftentimes the first few principal components explain the majority of the variability in the data. Therefore, PCA allows a researcher to deal with the problem in a much reduced dimension. This provides the basis for the PVCA procedure, and as its first step, the first few principal components are chosen based on one or more criteria such as a predefined threshold of the percentage of variability captured by them. Secondly, PVCA leverages the intuitive classical variance component estimation procedure by using all factors and interaction terms of interest as random effects, and fits them separately to each of the chosen principal components as a univariate response. In each mixed model fitting, REML provides an efficient and reasonable estimation method for each variance component. Fitting the mixed model to principal components, which themselves are spatially orthogonal, ensures that all the estimated variance components can be summarized using their corresponding eigenvalues as weights. In the end, all the potential sources of variability can be standardized to the sum of all variance components (including the residual variance, unexplained by the chosen effects in the model). A pie chart or a bar chart of the proportions provides a useful illustration of the partitioning of the total variance in the data.

As shown by the use case scenarios on three publicly available microarray data sets, the PVCA algorithm was capable of providing an estimate of the contribution from each factor of interest in terms of variance. This procedure will also provide a reasonable way to inspect the changing variability contribution of the corresponding effects before and after normalization. For an illustration of this, see Figure 13.1(c) in the next chapter. On the lung cancer data, it extracted two major sources of variability equivalent to 87.72% of the total variability: one being the batch effect contributing almost half of the total, and the other directly relevant to the experimental hypothesis. After empirical Bayes (ComBat) batch correction procedure proposed by Johnson *et al.*, PVCA also revealed that batch effect only contributed 1% of the total variability on a subset of 752 genes, confirming the major conclusion from Johnson *et al.* (2007).

In the analysis results from the NIEHS toxicogenomics study, although no obvious batch effect was detected, the algorithm showed the proportion of variability those factors and interaction terms possessed and how they quantitatively relate to unexplained variability. We investigated the scenario from a slightly different angle. Without the chemical effects, the total variability was shared by dye effect, time, treatment and dose level. Interestingly, the results revealed that the experimental rats showed more prominent responses to N-nitrosomorpholine and thioacetamide exposure, which agreed with previous toxicological studies using these compounds (data not shown).

The assumptions and features of PCA provide the foundation for the PVCA procedure. The variance–covariance or correlation coefficient matrix is computed solely based on the mean and variance. However, among all distribution families, only the exponential distribution family (Gaussian, exponential, etc) can be described well by the first two moments (mean and standard deviation). The major assumption that all PCA procedures are based on is that the larger the variances that an axis possesses, the more important that axis will be. This may not hold true at all time. In fact, oftentimes when systematic (nonbiological) noise exists in an experiment and plays a major role in contributing to the total variability of the data, one will not obtain a meaningful interpretation of the data. Only when the first eigengene or the first few eigengenes are filtered out can the researcher see the real biological interpretation (Alter *et al.* 2000). Fortunately, we are using PCA as part of our effort to detect batch correction and violation of this assumption does not seem to cause any concern. In general, caution needs to be exercised when interpreting the PCA results to avoid misinterpretations. In PVCA, we have set the total variance explained by the selected principal components to pass a threshold. This usually ensures that the majority of the variability has been included in the VCA and helps with computation speed (which is usually fast), although one can include as many variance components as desired for particular applications.

Besides providing the overall quantified variability for effects, it is also desired to inspect the corresponding variability for each principal component during the PVCA process. Often, certain effect(s) may be more highly associated with certain principal component(s) and this phenomenon can be easily identified by inspecting the corresponding variability of effects for each principal component. Conceptually, if there is only major variability associated with batch effect for certain principal component(s), adjusting the corresponding principal component(s) from data may be another method of batch effect removal. This approach is similar to the singular value decomposition approach, but with guidance from PVCA for specifically removing the principal component(s) associated with batch effect.

PVCA is implemented in the Correlation and Principal Components process in the JMP Genomics software available from SAS. In addition, R and SAS source code is available at http://www.niehs.nih.gov/research/resources/software/pvca.

13

Batch Profile Estimation, Correction, and Scoring

Tzu-Ming Chu, Wenjun Bao, Russell S Thomas, and Russell D Wolfinger

Abstract

Batch effects increase the variation among expression data and, hence, reduce the statistical power for investigating biological effects. When the proportion of variation associated with the batch effect is high, it is desirable to remove the batch effect from data. A robust batch effect removal method should be easily applicable to new batches. This chapter discusses a simple but robust grouped-batch-profile (GBP) normalization method that includes three steps: batch profile estimation, correction, and scoring. Genes with similar expression patterns across batches are grouped. The method assumes the availability of control samples in each batch, and the corresponding batch profile of each group is estimated by analysis of variance. Batch correction and scoring are based on the estimated profiles. A mouse lung tumorigenicity data set is used to illustrate GBP normalization through cross-validation on 84 predictive models. On average, cross-validated predictive accuracy increases significantly after GBP normalization.

13.1 Introduction

Since microarrays were introduced more than a decade ago (Schena *et al.* 1995), this technology has become a staple in genomics research. Requests for genomics data review in new drug submissions or applications to the Food and Drug Administration have been steadily increasing since 2004 (Frueh 2006). As the technology has begun to take on a larger role in drug development, concerns have grown about data quality and its impact on analysis results. To address these concerns, the MicroArray Quality Control (MAQC) project was initiated by the National Center for Toxicological Research (NCTR) in 2005. A series of articles related to microarray quality control were published by the studies in

Batch Effects and Noise in Microarray Experiments: Sources and Solutions edited by A. Scherer
© 2009 John Wiley & Sons, Ltd

the first phase of the MAQC project in 2006 (MAQC Consortium 2006). Currently, the second phase of the MAQC project is focusing on predictive modeling and has submitted another series of articles likely to be published in 2009.

Normalization is typically the first step in microarray data analysis and has a large impact on downstream analysis. The goal of normalization is to correct for unwanted nonbiological effects, such as background surface effect, hybridization-wash efficiency, and cDNA labeling efficiency across arrays. Scores of normalization methods have been proposed, including analysis of variance (ANOVA)-based median or mean normalization (Kerr and Churchill 2001a; Wolfinger *et al.* 2001), loess normalization (Yang *et al.* 2002b), invariant-gene-set normalization (Schadt *et al.* 2001), and quantile normalization (Irizarry *et al.* 2003). All of these techniques generally have a positive effect by increasing correlations among arrays and enhancing the statistical power to identify significant genes with differential expression across treatment groups. Bolstad *et al.* (2003) and Rao *et al.* (2008) compare different normalization methods with an assessment of correlations among arrays and differential expression of spiked-in genes. However, batch effects were not examined or taken into account in these studies, and the degree of normalization is closely associated with data quality.

In practice, certain limitations, such as the capacity for sample processing, different vehicles for transferring component treatments, and handling of repeated samples from different time periods, often restrict the corresponding microarray experiment to being conducted in batches. In such cases, the batch effects play a significant role in the overall variability within a data set that may not be reduced using standard normalization methods. Failing to account for significant batch effects will typically increase variability and reduce the statistical power of downstream data analyses.

Scatterplots of principal components are a handy tool for visualizing batch effects. Variance components analysis of principal components (see Chapter 12 in this volume) can further estimate the relative proportion of variation associated with the corresponding batch effect. Figure 13.1 shows an example of visually examining batch effects in a principal component analysis (PCA) plot and the corresponding variance component plot. When the corresponding proportion of variation associated with batch effects is high, it is desirable to remove the batch effect from data.

Several different approaches for correcting batch effects have been introduced, including the use of singular value decomposition (SVD; Alter *et al.* 2000), distance weighted discrimination (DWD; Benito *et al.* 2004), and empirical Bayes (Johnson *et al.* 2007). SVD normalization depends on the selection of the first several eigenvectors associated with the batch effect and their subtraction from the data. Determining which eigenvectors to use in SVD normalization can be a challenge, and there is a definite risk of over-correction or removal of biological signal. DWD is a modified version of support vector machines and can be applied for normalizing two batches at a time. A stepwise approach illustrated in Benito *et al.* (2004) can work with multiple batches with a similar strategy to hierarchical clustering. It starts by normalizing the two most similar batches and then considers them as one batch. In the next step, the previous joint one and the third batch are DWD-normalized. These steps are repeated until all batches are DWD-normalized. Different ordering for batches in stepwise DWD normalization may lead to different normalized results. Also, batches may have low similarity and increase the difficulty of using stepwise DWD. While SVD and DWD normalization usually require relatively

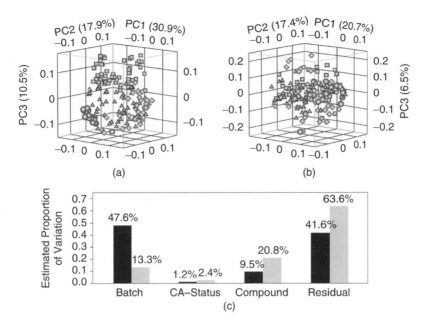

Figure 13.1 (a) PCA plot before batch normalization. (b) PCA plot after GBP normalization. (c) Bar chart of estimated proportions of variation. The circle, triangle, diamond, and square symbols in (a) and (b) indicate the array data from studies 0503, 06004, 06014, and 07008, respectively (see text for further details). The black and gray colors in (c) indicate the estimated proportion of variation before and after normalization, respectively.

large sample sizes, the empirical Bayes approach with shrinking variance components (Lönnstedt and Speed 2002; Feng *et al.* 2006) can be a robust approach for smaller sample sizes. The empirical Bayes batch normalization in Johnson *et al.* (2007) can be appealing, but there is no scoring feature for normalizing future batches.

When collecting microarray data for predicting an outcome of interest, such as tumor incidence rate and cancer status, the ability to score and adjust future samples is essential for a robust normalization since similar batch effects may be present in these samples. A robust batch effect removal method should be easy to apply to new batches without renormalizing the entire data set. This chapter will discuss a simple but robust grouped-batch-profile (GBP) normalization method that includes three steps: batch profile estimation, correction, and scoring. Genes with similar expression across batches are grouped. The corresponding batch profile of each group is estimated by an ANOVA model, and then batch correction and scoring are based on the estimated profiles. A mouse lung tumorigenicity data set is used to illustrate GBP normalization through cross-validation on 84 predictive models.

13.2 Mouse Lung Tumorigenicity Data Set with Batch Effects

The purpose of the studies that comprise the mouse lung tumorigenicity data set was to evaluate the feasibility of developing microarray-based biomarkers that can predict the results from a rodent cancer bioassay following a relatively short-term chemical exposure. Currently, rodent cancer bioassay is used as the primary means to determine the carcinogenic potential of a chemical and generate quantitative information on the dose-response behavior for chemical risk assessments. The bioassays are expensive, time-consuming, and, according to the National Toxicology Program, use hundreds of animals (NTP 1996). Fewer than 1500 chemicals have been tested in a rodent cancer bioassay compared to the thousands of environmental and industrial chemicals that remain untested for carcinogenic activity (Gold *et al.* 1999).

The data for the mouse lung tumorigenicity study can be obtained from http://www.ncbi.nlm.nih.gov, GSE17933. The study design involved selecting 26 diverse chemicals from those tested by the NTP. Approximately half were positive for increased lung tumor incidence in female B6C3F1 mice and half were negative. Female mice were exposed for 13 weeks to each of the chemicals and microarray analysis was performed on the lung. Microarray measurements were conducted on three to five mice per treatment group. Female B6C3F1 mice were chosen since they represent the most sensitive model for lung tumors in NTP studies. The chemical exposures and microarray hybridizations were performed as a series of four studies over a period of 3 years (Table 13.1, Animal Study ID). Some studies had staggered starts with different batches of animals (Table 13.1, Animal Batch). Each chemical was run with a concurrent vehicle control group. In total, 158 Affymetrix Mouse Genome 420 2.0 microarrays were analyzed in the data.

The Affymetrix array data was extracted from probe-level CEL files and initially normalized by subtracting the within-array probe mean for each array after summarizing to probe means for each probe set. These pre-processed data will be referred to as mean-mean adjusted (MMA) data in this chapter. They are used as the input data for

Table 13.1 Batch groups with corresponding compounds for the mouse tumorigenicity data.

Batch	Animal study ID	Animal batch	Vehicle group	Controls	CA	NCA	Compounds
1	5003	1	FCON	1	1	2	4
2	5003	1	CCON	1	1	0	2
3	6004	1	FCO2	1	1	3	5
4	6004	2	CCO2	1	2	0	3
5	6004	3	CCO3	1	1	0	2
6	6004	4	WCO1	1	1	1	3
7	6014	1	CCO4	1	0	2	3
8	6014	2	FCO1	1	2	0	3
9	6014	3	ACO1	1	0	2	3
10	7008	1	CCO5	1	0	1	2
11	7008	2	ACO3	1	1	1	3
12	7008	3	CCO6	1	1	0	2
13	7008	4	ACO4	1	2	0	3
14	7008	5	ACO5	1	1	0	2

Note: The first three letters in the abbreviation define the vehicle and route used for administration (FCO = food; CCO = corn oil by gavage; WCO = water by gavage; ACO = air by inhalation). The last number or letter refers to the specific batch in which the vehicle was used.

GBP normalization. For comparison purposes, RMA-normalized data were also generated. Since the major purpose of this mouse lung tumorigenicity data is to predict the carcinogenic potential of future chemical compounds, the 158 array data set was averaged across replicate arrays within chemical group. The corresponding number of compounds within each batch is listed in Table 13.1.

Figures 13.1(a) and 13.1(b) show the PCA plots for the MMA and GBP normalized data. Visually, the batch effect is highly correlated to the four studies in the MMA data. Figure 13.1(c) shows the estimated proportion of variation for both data sets with over 90% of accumulated proportion of variation. The proportion of variation for batch effect is dramatically reduced from 47.6% to 13.3% after GBP normalization. See Chapter 12 in the present volume for details of how to estimate proportion of variation.

13.2.1 Batch Profile Estimation

Batch profile estimation is the first step of GBP normalization. If we consider the observed $\log_2$ intensity for a transcript to be additively described by biological signal, batch effect, and stochastic error, the following model is reasonable:

$$Y = \alpha(s) + \beta(b) + e, \tag{13.1}$$

where Y represents observed $\log_2$ intensity, α is a linear function representing biological signal which usually consists of fixed and random effects, β is a function representing batch profile, and e is the corresponding error term. Assuming that the average (or overall) biological signal within each batch is similar, the variation of average observed intensity within each batch should be highly related to the corresponding batch effect. Without loss of generality, this assumption can be valid for a balanced design with all batches sharing the same combination of treatments.

The use of vehicle controls for delivering chemical compounds is a common practice in toxicity experiments. The control vehicle arrays are very good candidates for estimating batch profiles. The previous formula can be adjusted for controls as follows:

$$Y_c = \alpha(c) + \beta(b) + e, \tag{13.2}$$

where Y_c is observed $\log_2$ intensity in a control array and $\alpha(c)$ is a constant baseline expression for all the controls. For simplicity, assuming $\beta(b)$ has a single additive term for a single batch level,

$$Y_{ci} = \mu_c + \beta_i + e, \tag{13.3}$$

where Y_{ci} is the observed $\log_2$ intensity in a control array in batch i, μ_c is the constant baseline $\log_2$ intensity for all controls across all batches, and β_i is the batch effect for batch i. Therefore, the batch profiles β can be estimated by the simple one-way ANOVA in (13.3) with

$$\hat{\beta}_i = \overline{Y}_{ci} - \overline{\overline{Y}}_{ci}, \tag{13.4}$$

where

$$\overline{Y}_{ci} = \frac{1}{n_{ci}} \sum_{\text{arrays in batch:}} Y_{ci}, \quad \overline{\overline{Y}}_{ci} = \frac{1}{n_c} \sum_c \overline{Y}_{ci},$$

in which n_{ci} is the number of control arrays in batch i, and n_c is the number of batches. $\overline{Y}_{ci}$ is the within-batch mean of the controls. $\overline{\overline{Y}}_{ci}$ is the cross-batch mean of the controls and the estimator of μ_c. More sophisticated functions can potentially be used for the batch profile function $\beta(b)$ in (13.1) and (13.2). When applying a location and scale (LS) adjustment, this will be similar to the model applied in Johnson *et al.* (2007). Either applying only the location adjustment as in (13.3) or the LS adjustment can also be done using an empirical Bayes approach. For simplicity of scoring future batch profiles, we will only apply model (13.3) for the mouse lung tumorigenicity data in this chapter.

13.2.2   Batch Profile Correction

Once the batch profiles are estimated, batch correction can be done by simply subtracting them from observed $\log_2$ intensities as

$$\hat{Y} = Y - \hat{\beta}_i \qquad\qquad (13.5)$$

where $\hat{Y}$ is the batch-adjusted $\log_2$ intensity. This adjustment is performed on genes one by one.

Inspecting the estimated batch profiles reveals that some genes have highly correlated batch profiles. In order to group those genes with similar batch profile together, k-means clustering was carried out on standardized batch profiles, which are computed by subtracting the mean and dividing by the standard deviation of the corresponding batch profile. Figure 13.2 shows three clusters of standardized batch profiles from $500\, k$-means clusters. We have observed similar grouped batch profiles in several other microarray data sets. Note that the batch profiles in Figure 13.2 are derived from vehicle control arrays. Those vehicle controls are not expected to introduce biological variation in the data.

When there are multiple sources causing the batch effect, some genes can have high correlation across batches. This can be considered gene-by-batch interaction. In the other words, different groups of genes may behave very differently across batches. Taking this into account, the single batch correction in (13.5) can be updated for GBP normalization. The mean profile of each group is applied as the representative profile of the corresponding group. Therefore,

$$\tilde{Y} = Y - \widetilde{\beta}_{gi}, \qquad\qquad (13.6)$$

where $\tilde{Y}$ is GBP normalized $\log_2$ intensity and $\widetilde{\beta}_{gi}$ is the mean profile of the corresponding group g in batch i. When k-means clustering is performed on standardized batch profiles, the representative mean profile $\widetilde{\beta}_g$ is multiplied by the corresponding standard deviation to invert the standardization. Compared to single gene normalization, GBP gains a robust advantage by relying on highly correlated genes across multiple batches with adjustment from the majority of genes in the group if a single gene behaves differently in a batch but highly consistently with the majority of genes in the other batches.

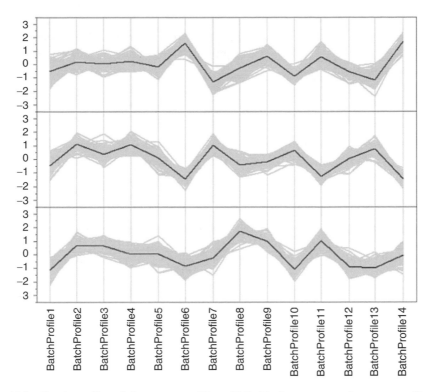

Figure 13.2   Batch profiles of three groups. The middle black curves are the mean profiles of the corresponding groups.

13.2.3   Batch Profile Scoring

Based on models (13.3) and (13.4), scoring for batch profiles in a new batch is straight-forward by subtracting the estimated constant baseline expression for all the controls. Since the baseline expression of controls is expected to be constant, the corresponding estimated value from previous GBP normalized data of the same experiments can be saved and applied. Therefore,

$$\hat{\beta}_j = \overline{Y}_{cj} - \overline{\overline{Y}}_{ci}. \tag{13.7}$$

Here, indexes i and j indicate old and new batches. With the new scored batch profiles, GBP normalization can be applied to new batches. Based on the groups defined in previous data, the scored profiles are grouped and the corresponding representative mean profiles are calculated. In order to avoid influential outliers when calculating the mean, some robust approach such as trimmed mean, which skips observations at both low and high ends, and winsorized mean, which replaces the observations at both low and high ends with the closest non-winsorized observations, can be applied here. For the example in this chapter, the winsorized approach is applied with 5% rate (2.5% of observations in each tail). Once the representative mean profiles are estimated, model (13.6) is applied for GBP normalization.

13.2.4 Cross-Validation Results

To investigate the practical usefulness of GBP normalization, the mouse lung tumorigenicity data are used with cross-validation on a set of 84 predictive models. These 84 models are selected with different parameters for eight types of predictive model implemented in the JMP Genomics software (http://www.jmp.com/software/genomics/). The eight types of predictive models are discriminant analysis, distance scoring, generalized linear model, K nearest neighbors, logistic regression, partial least squares, partition trees, and radial basis machine. The corresponding parameters are listed in Tables S13.1–S13.9 on the book website. Refer to Tables S13.1–S13.8 for the SAS macro names applied and Table S13.9 for the corresponding JMP Genomics parameter names. All the parameters not included here are the same as the corresponding defaults in JMP Genomics 4.0. GBP normalization is also implemented in JMP Genomics 4.0. Trial version of JMP Genomics can be requested from http://www.sas.com/apps/forms/index.jsp?id=genjmp_genomics. Similar routines are available in R or Matlab.

For cross-validation, three batches from the entire 14 batches in Table 13.1 are randomly selected as a hold-out validation data set and the remaining 11 batches are used as a training data set. Both the training and validation data sets are further processed by averaging across arrays within compound group before predictive modeling. The 84 models are fitted on the training data set and the corresponding fitted models are applied to predict the hold-out validation data set. The splitting, modeling, and scoring processes are repeated 50 times. The three normalized data sets from MMA, RMA, and GBP are compared. Figure 13.3 illustrates these processes by means of flowcharts. MMA and RMA normalization are performed once before splitting training and validation data sets, whereas the GBP normalization is repeatedly performed after splitting training and validation data sets prior to training models. Three statistics, accuracy, area under receiver operating characteristics curve (AUC), and root mean square error (RMSE), are applied for assessment.

Figure 13.4 shows the cross-validation results. The circle, plus, and diamond symbols in Figure 13.4 represent the results from MMA, RMA, and GBP normalized data, respectively. Almost all of the GBP results are superior to those of MMA and RMA across all 84 models, with the diamond line higher than the other two lines for accuracy and AUC and lower than the other two lines for RMSE. This is evidence that GBP normalization leads to better predictive outcomes with lower error. A few exceptions occur with AUC values in some of the partition tree models on MMA data. The models of partition trees are numbered from 65 to 80 in Figure 13.4. Averaging across the 84 models, the results of accuracy for applying MMA, RMA, and GBP normalized data are 0.56, 0.54, and 0.66, respectively. The corresponding average triplets for AUC and RMSE are (0.59, 0.57, 0.69) and (0.58, 0.59, 0.51), respectively. The simple MMA normalization exhibits a slight 0.02

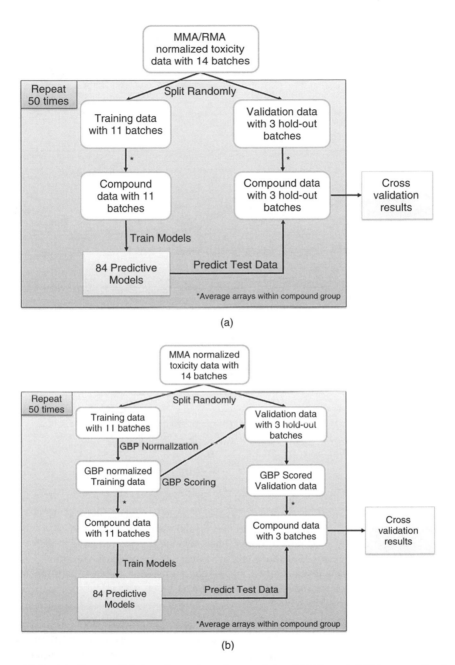

(a)

(b)

Figure 13.3 (a) Cross-validation flow chart for applying MMA and RMA normalized data. (b) Cross-validation flow chart for applying GBP normalized data. MMA and RMA normalization are performed once, whereas the GBP normalization is repeatedly applied after splitting training and validation data sets. All data sets are averaged across arrays within each compound group prior to training models. (Color version available as Figure S13.3 on website.)

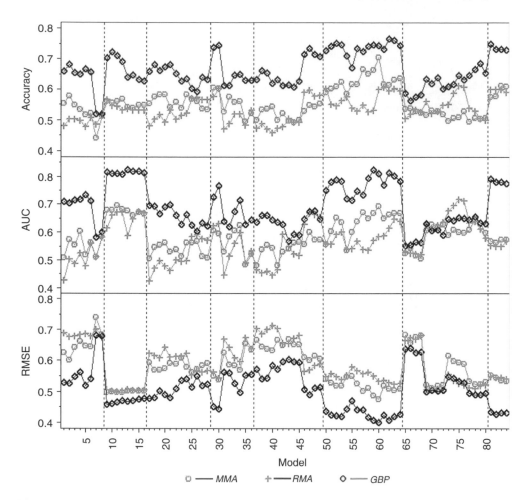

Figure 13.4 Cross-validation results of the 84 predictive models. The three plots from top to the bottom are the cross-validation results of comparing accuracy, AUC, and RMSE, respectively. The circle, plus, and diamond symbols indicate the results with applying MMA, RMA, and GBP normalizations, respectively. The vertical dotted lines separate the 84 predictive models into eight types in the following order from left to right: discriminant analysis, distance scoring, generalized linear model, K nearest neighbors, logistic regression, partial least squares, partition trees, and radial basis machine. (Color version available as Figure S13.4 on website.)

improvement over RMA normalization in both accuracy and AUC assessments. GBP is then 0.1 higher than MMA normalization in both accuracy and AUC assessments.

13.3 Discussion

Batch effects can negatively impact microarray analysis results, especially for data collected over multiple years. Viewing data with PCA plots as in Figures 13.1(a) and 13.1(b) is useful for visually inspecting batch effects. Calculating the corresponding proportion of variation through variance component analysis of principal components as in Figure 13.1(c) can quantitatively reveal the degree of batch effect with respect to total variation.

Batch profiles can be easily updated using equation (13.4). The first term on the left-hand side of (13.4), which is the within-batch control mean, is fixed once the corresponding data are collected. The second term, the grand mean of within-batch control means across batches, can be updated by recalculating it when adding new batches. This gives ability for the GBP to be fine-tuned as new data become available. The corresponding groups of batch profiles can be updated with a new run of k-means clustering. The number of groups of batch profiles, defined as the number of clusters from k-means clustering, is a tuning parameter in GBP normalization. Instead of fixing a specific number of clusters, one can instead set a maximum correlation radius distance and let the number of clusters be determined by it.

When batch effects are present, some form of batch scoring is essential when applying microarray data for prediction. It should be possible and easy to apply normalization to new data for scoring prediction. Assuming the availability of control samples in each batch, GBP normalization is straightforward and includes scoring functionality that is easily incorporated in the prediction framework. Our 'leave-three-batches-out' cross-validation example exhibits prediction improvements gained by GBP normalization over MMA and RMA normalization.

Acknowledgements

The authors thank SAS colleagues Glenn Horton and Stan Martin for help with setting up and running SAS Grid Manager to parallelize the cross-validation computations.

The microarray studies that comprise the mouse lung tumorigenicity data set were supported by the American Chemistry Council's Long Range Research Initiative.

14

Visualization of Cross-Platform Microarray Normalization

Xuxin Liu, Joel Parker, Cheng Fan, Charles M Perou, and J S Marron

Abstract

Combining different microarray data sets, even across platforms, is considered in this chapter. The larger sample sizes created in this way have the potential to generally increase statistical power. Distance weighted discrimination (DWD) has been shown to provide this improvement in some cases. We replicate earlier results indicating that DWD provides an effective approach to cross-platform batch adjustment, using both novel and conventional visualization methods. Improved statistical power from combining data is demonstrated for a new DWD based hypothesis test. This result appears to contradict a number of earlier results, which suggested that such data combination is not possible. The contradiction is resolved by understanding the differences between gene-by-gene analysis and our more complete and insightful multivariate approach of DWD.

14.1   Introduction

DNA microarrays have proven to be a powerful tool for many biological applications. But serious statistical challenges remain because of biological and technical variation in the data. This variation could be countered by running a large number of arrays and averaging the results, but this is currently not practical because array costs are still relatively high. Another approach to boosting statistical power is to combine current data with previously collected data, much of which is available on the web.

Hurdles to such combinations include biases introduced during the sample preparation, manufacture of the arrays, and the processing of the arrays (labeling, hybridization, and scanning). Even more challenging is that the data can seem especially noncomparable when they are collected using different platforms (Affymetrix, Agilent, and others). Distance weighted discrimination (DWD), developed by Marron and Todd (2002) and Marron

Batch Effects and Noise in Microarray Experiments: Sources and Solutions edited by A. Scherer
© 2009 John Wiley & Sons, Ltd

et al. (2007), was shown to provide effective bias adjustment for all of these situations by Benito *et al.* (2004), where it was also seen to be effective for cross-platform adjustment.

Despite these promising results, there have been a number of contradictory results, suggesting that these systematic biases are an insurmountable obstacle for cross-platform analyses (see Kuo *et al.* 2002; Parmigiani *et al.* 2004; Mecham *et al.* 2004). The first of these publications is based on an unusually direct comparison of the National Cancer Institute (NCI) 60 cancer cell line data. These data provide an excellent test bed for studying cross-platform issues because gene expression of identical cell line samples was measured by both cDNA (from Synteni, Inc.; now Incyte, Inc.) and Affymetrix (Hu6800) microarrays. The expression values are available on the internet at http://discover.nci.nih.gov/datasetsNature2000.jsp and http://discover.nci.nih.gov/datasetsPnas2001.jsp. Because the samples are identical, the effectiveness of cross-platform adjustment can be precisely calibrated. Finally, because the cDNA expression values are on the scale of differences of log intensities (without commonly used loess normalization in this cDNA data set), we also work with the logarithm to base 2 of the Affymetrix data (the original data were generated by Affymetrix Microarray Suite 4.0). The cDNA values had some missing data points, which we handled by imputation using K-nearest-neighbor imputation (see Troyanskaya *et al.* 2001).

The above discussion assumes that genes have been appropriately matched across platforms. We recommend using Entrez Gene IDs for the purpose, because they are relatively stable over time. Another choice is to use UniGene Cluster IDs. See Culhane *et al.* (2003) for an interesting approach to poorly matched gene sets.

This NCI60 test bed is used to investigate the effectiveness of DWD in Section 14.2. It is seen that DWD adjustment allows combining the cDNA and Affymetrix data sets into a single homogeneous data set, which contains all the previously known biological features of these data. Multivariate projection views are quite important both to the approach and to understanding the adjustment process.

While the visualizations in Section 14.2 strongly suggest that the DWD cross-platform adjustment is successful, it does not directly settle the central issue: does the adjustment allow the combined data to have improved statistical power? This important question is addressed in Section 14.3, where hypothesis tests are considered to study whether each cancer type is statistically significantly different from the rest of the data. It is seen that in almost every case, the adjusted data provide improved statistical power, thus justifying the combination of data.

While Benito *et al.* (2004) – and a lot of unpublished experience in our lab – showed that DWD is superior to most methods of data normalization (e.g. centroid based) in practice, an interesting question is why this happened. This has been conclusively answered by the results of Liu (2007), who showed using an intuitive example (and validated using asymptotic mathematical statistics) that the phenomenon is driven by unbalanced subclasses within each platform. Essentially the sample means (and other least squares based approaches) are strongly affected by the relative proportions of subclasses, while DWD feels the data in a way that leaves them very robust against this type of unbalancing.

Resolution of the apparent contradiction between the clearly positive results of Sections 14.2 and 14.3 and some earlier negative results, using the same data, comes in Section 14.4. There a simulated example is used that shows that important biological structure can be missed by restricting attention to a gene-by-gene view. Furthermore, it is seen that a gene-by-gene analysis of correlation can suggest negligible correlation between two sets of samples. Yet after DWD adjustment, the same samples can give an extremely high multivariate correlation, in the direction of biological interest.

An important caveat to the application of DWD is that all important biological classes need to be represented in both subgroups to be adjusted. That is to say, if biological type 1 lies completely in the first group, and type 2 lies completely in the second group, and DWD is applied to these data, then DWD will eliminate the differences between groups, which means that it will also eliminate the important biological differences in the process.

14.2 Analysis of the NCI 60 Data

Figure 14.1 studies the NCI 60 cancer cell line data, using a visualization that will be used frequently in this chapter. The most important property of this view is that it is multivariate in nature, as opposed to more conventional gene-by-gene views, such as is commonly done when using conventional heatmaps (i.e. hierarchical clustering diagrams). The challenge to multivariate data views is that the human perceptual system is only capable of understanding one-, two-, or three-dimensional views. An approach to this issue is to focus attention on one- and two-dimensional projections, based on carefully chosen directions of interest, and chosen from the many that are possible to consider. Principal component analysis (PCA) is a commonly used method for finding directions of interest. This gives a set of multivariate directions which are orthogonal to each other and frequently provide useful views because these are the directions that maximize the spread of the projected data. But other directions can be very useful as well, particularly to highlight known differences of various types in the data. An example of this is shown in the step-by-step illustration of the DWD batch and source adjustment, available from the 'DWD Bias Adjustment of Batch and Source Effects' link at http://genome.med.unc.edu:8080/caBIG/DWDWebPage/DWDNCI60.htm. In particular, while principal component (PC) directions are often useful, for some purposes it is very insightful to include DWD direction vectors as well.

In Figure 14.1, the chosen directions are the first four PC axes, where the PCs have been computed using the full data set. The plots on the diagonal show the one-dimensional projections of the data. The cDNA observations appear as black plus signs (each plus sign represents one sample, i.e. array), and the Affymetrix data are gray circles. The axis shows the projections of the data on each PC direction vector. A random height is added to each symbol just for convenient visual separation – essentially the 'jitter plot' idea of Tukey and Tukey (1990). Also included in each plot is a smooth histogram, colored according to the microarray platform. Note that PC 1 points essentially in the direction of the platform difference, because this is the direction of greatest variation in the combined data set. The off-diagonal plots are projections of the data onto two-dimensional planes, determined by the various pairs of the PC directions. These are scatterplots where the horizontal axis shows the one-dimensional projection that appears in the same column, and the vertical axis shows the one-dimensional projection in the same row. Again symbols correspond to samples, and the symbol type and shade of gray indicate the platform. Also present in these scatterplots are line segments connecting the samples from common cell lines recalling that the same cell lines were assayed on each platform. See the Scatter Plot View of Micro-Array Data web

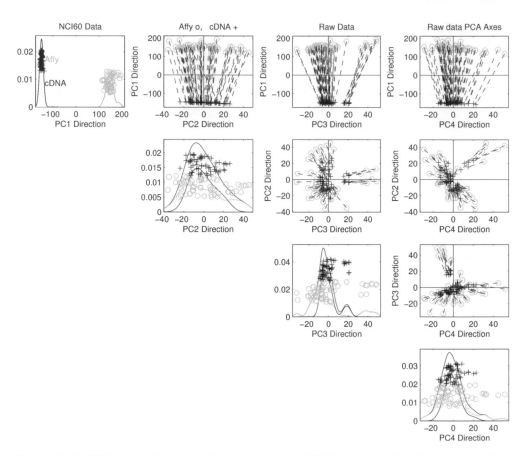

Figure 14.1 PCA scatterplot view of raw (log scale) NCI 60 data. Dashed lines connect identical samples. This shows a large measurement differences between the two platforms, so some
adjustment is essential before combining data sets. Yet differences are systematic, which offers
hope of careful adjustment. For this, and also all following figures, more detailed explanation of
the graphics can be found in the text. The vertical axes in the plots on the diagonal (here and in
similar plots throughout the chapter) are not labeled because they play two roles: first in units of
'density per unit area' for the smoothed histogram, second in units of 'order in data set' for the
jitter plot. (Color version available as Figure S14.1 on website.)

page (http://genome.med.unc.edu:8080/caBIG/DWDWebPage/ScatPlotDataView.htm) for
a detailed, step-by-step introduction to this type of graphic. The other PC directions show
other types of structure, which will be discussed in detail below.

Already apparent in Figure 14.1 is some suggestion of biological clusters. For example
in the PC2 versus PC3 scatterplot (the second panel in the second row) there is a cluster
that seems to separate itself from the rest. However, the cluster is not very distinct in
the sense that the distance between the two platforms is as large as the separation of the
cluster from the main body of data. Another potential cluster seems to appear in the PC2
versus PC4 scatterplot (the last panel in the second row), but again the cross-platform
distances are very large relative to the cluster separation from the main data.

Perfect cross-platform adjustment would change the data in such a way that each cDNA sample (plus) would coincide with its corresponding Affymetrix sample (gray circle), and each connecting line would have a length of 0. Of course this is impossible because these measurements were made in the presence of technical variation. However, in all of the off diagonal panels visible in Figure 14.1, the line segments do follow intriguingly simple patterns, suggesting that in fact some relatively simple operations could yield considerable overlap of the desired type. DWD and some subsequent adjustment steps, are aimed at accomplishing this goal. Most of the figures in this chapter are available in color versions on the book's website. They are much easier to interpret. For the purpose of this print version of the book, gray levels are used instead.

Figure 14.2 shows (using the same view) the same data after DWD adjustment. This adjustment used both DWD to find the right direction for shifting the data, followed by a columnwise standardization, which is important to correctly handle scale differences present across platforms. DWD alone would be sufficient if the connecting lines were all parallel. A careful, step-by-step visual display of the steps in this adjustment is available on the DWD Cross-Platform Adjustment of the NCI-60 Data web page (http://genome.med.unc.edu:8080/caBIG/DWDWebPage/DWDNCI60.htm). Note that, in all of the PC directions, the huge difference between the cDNA and Affymetrix data visible in Figure 14.1 has essentially disappeared. The dashed line segments, which connect samples from the same cell line, show that there are both some systematic differences, in the sense that many of the nearby line segments are approximately parallel, and some pure noise differences, reflected by nearby line segments lying in much different directions. But the key observation is that both types of noise are smaller in magnitude than the distinct clusters that are visible in the data. These clusters represent important biological structure in the data (as detailed below), which shows that the DWD normalization has reduced differences in the data to a level which is less than the biological features in this data. This is the key to effective combination of data from across different statistical platforms.

The biological significance of the clusters visible in the combined data is studied in Figure 14.3. This is the same as Figure 14.2, except that now the two clearly visible clusters are colored. There is a medium gray colored cluster, which shows up clearly in all of the PC 3 views (third column and third row), and a light gray cluster, which shows up clearly in the PC 2 versus PC 4 scatterplot in the last column and the second row. The names of the cDNA arrays for the data in these clusters are shown in Table 14.1.

The samples in the medium cluster (left-hand column) in Table 14.1 are all melanoma cell lines except two, which was previously shown by Ross et al. (2000), using hierarchical clustering analysis, to be a very strong cluster that is very noticeably different from the other cancer types. Also note that two breast cancer cell lines also appear in this cluster, which again repeats the previous observations of Ross et al. The points in the light gray cluster are all leukemia cell lines that are derived from blood lymphocytes. This is a second dominant expression pattern that reflects cell type identity, and was also identified by Ross et al. Both of these clusters are further studied using conventional heatmap views; see these on the DWD Cross-Platform Adjustment of the NCI-60 Data web page.

A much different view of the DWD adjusted NCI 60 data, which particularly focuses on the known biological clusters, is shown in Figure 14.4. The scatterplots shown in Figure 14.3 are informative, but the directions used in the projection view are PCA

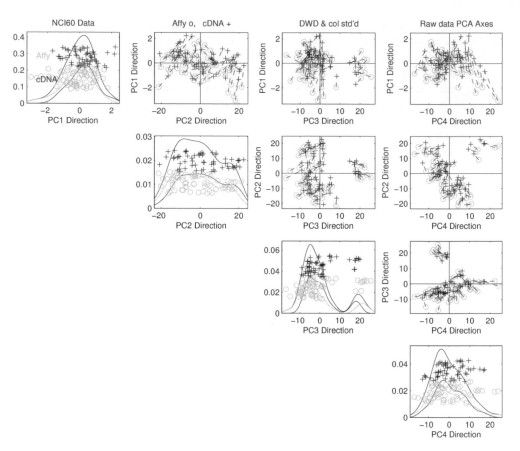

Figure 14.2   PCA scatterplot view of DWD adjusted NCI 60 data, showing effective removal of platform bias. In particular, distances between same cell lines (shown as dashed lines) are now much smaller than distances between apparent clusters. (Color version available as Figure S14.2 on website.)

directions, which are attuned to 'maximal variation (in the projected data)'. This direction frequently correlates well with important biological insights, but does not do so explicitly. While the melanoma and leukemia clusters appear quite clearly, the other cancer types are not easy to see, even when more PCs are studied. A completely different application of the DWD direction vector (from providing the key to bias adjustment as done above), is to provide directions that more directly target biological interest, as is done in Figure 14.4. The directions used there were computed by grouping the eight biological subtypes into pairs, as shown in the axis labels for each panel. For each pair DWD was used to find the direction vectors aimed at separating the two classes from each other.

Figure 14.4 shows that DWD was generally very successful in providing directions which drew strong distinctions between most biological classes and the remaining data. Note that for most classes there are considerable gaps between those clusters and the main body of the data. Not surprisingly, the melanoma (left-hand side of the plots in the third column) and leukemia (right-hand side of the plots in the fourth column) clusters

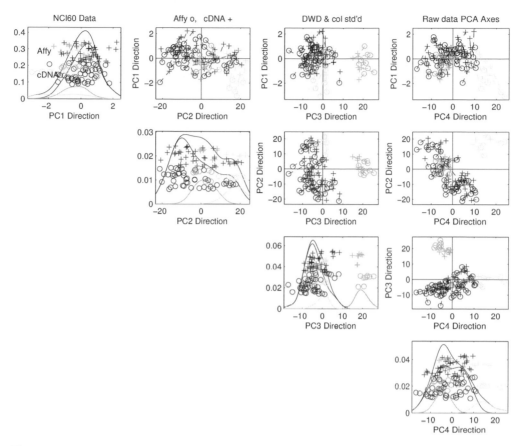

Figure 14.3 PCA scatterplot view of DWD adjusted NCI 60 data. Important biological clusters are highlighted: melanoma (medium gray), leukemia (light gray). (Color version available as Figure S14.3 on website.)

have the largest such gaps. Two exceptions to this are the non-small cell lung cancers (right-hand side of the plots in the second column) and breast cancer (left-hand side of the plots in the second column). This is likely due to the biological heterogeneity present in these groups, which, for the example of the breast cancer cell lines, contains lines with luminal and fibroblast-like characteristics (see Ross and Perou 2001).

The main result of Figure 14.4 is that for all of the biological classes, the differences between platforms (shown as connecting lines between the pairs of symbols representing each common sample) are much smaller than the biological differences between the biological classes. Thus it is not surprising that in Section 14.3 it will be seen that combining data produces improved statistical power (sensitivity).

Note that the axes shown in Figure 14.4 are not orthogonal to each other, unlike for the PCA based views shown above. For this reason, plots below the diagonal are also included, because they are different (while for the PCA directions the plots were just transposed, which added no new information, and hence were not included).

Table 14.1   Sample names in the two highlighted clusters. (Color version available as Table S14.1 on website.)

Medium gray cluster	Light gray cluster
BREAST.MDAMB435	LEUK.CCRFCEM
BREAST.MDN	LEUK.K562
MELAN.MALME3M	LEUK.MOLT4
MELAN.SKMEL2	LEUK.HL60
MELAN.SKMEL5	LEUK.RPMI8266
MELAN.SKMEL28	LEUK.SR
MELAN.M14	
MELAN.UACC62	
MELAN.UACC257	

While DWD has been applied to the raw data here, to best highlight the full potential of the method, current practice in our lab is to first normalize each data set by median-centering each gene, and then apply DWD.

While this visualization builds a strong case that the DWD cross-platform adjustment has been successful, it still does not directly consider the question of chief concern: is there value added, in terms of statistical power, by combining these data sets using DWD? This question is answered affirmatively in the next section.

14.3   Improved Statistical Power

In this section the focus is on the statistical problem of understanding which of the biological subtypes are statistically significantly different from the rest of the data. Figure 14.4 suggests the melanoma and leukemia clusters are clearly distinct, and the non-small cell lung and breast cancer clusters are not. But what about the less clear-cut clusters? How can these ideas be quantified in terms of p-values?

The DWD direction is used once again, in a different way here. This time, for each class the DWD direction that best separates it from the rest of the data is computed. Statistical significance is computed by projecting the data onto the direction vector, and then computing a two-sample t statistic.

The lower left-hand panel of Figure 14.5 shows the names of the eight cell lines that were labeled as renal. The name shown in light gray, RENAL.SN12C, had an expression pattern that was much different from the others, possible due to a mislabeling of the cancer type, so it is not used in the analysis presented here. The upper left-hand panel shows the projected data (again using circles for Affymetrix and plus signs for cDNA), where the DWD direction for separating the renal data from all of the rest is used. Note that the renal data (highlighted in gray) are quite distinct from the other data. We assess statistical significance of the renal cluster in terms of the difference in means between the renal data and the rest. Thus, also shown are the values of the two sample t statistic for the combined data (top line of text), for the Affymetrix only data (bottom line of text), and for the cDNA only data (middle line of text). Note that the combined t statistic is

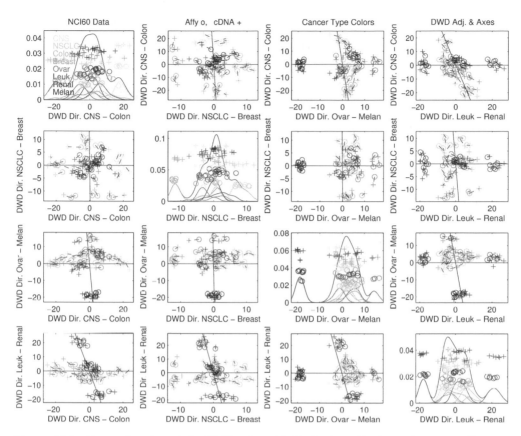

Figure 14.4 Scatterplot view of the DWD adjusted NCI-60 data, this time using DWD direction vectors. Different shades of gray indicate cancer classes. This shows that cross-platform differences are predominantly much smaller than differences between cancer types. (Color version available as Figure S14.4 on website.)

larger than the others, suggesting that the combined data have more statistical power than either individual platform. Another feature of this plot is that the Affymetrix t statistic is larger than for the cDNA, suggesting more statistical power for the Affymetrix data. It is important to resist the temptation to compare this number with the usual t distribution quantiles, because the DWD direction vector has a tendency to strongly magnify this statistic. While the t statistics contain some information about relative statistical power, this comparison is unfair to the individual platforms because the direction vector was chosen for the combined data, which could be different from the DWD direction vector for the individual samples. This issue is addressed in the center panels.

The top center panel of Figure 14.5 is similar to the top left-hand panel, except this time only cDNA data are considered. This is for computation of the DWD direction, for the projection, and for the computation of the t statistic. Note that the t statistic is now larger than the corresponding value in the top left-hand panel, which shows that indeed the comparison between combined and cDNA only tests is fairer when the DWD directions

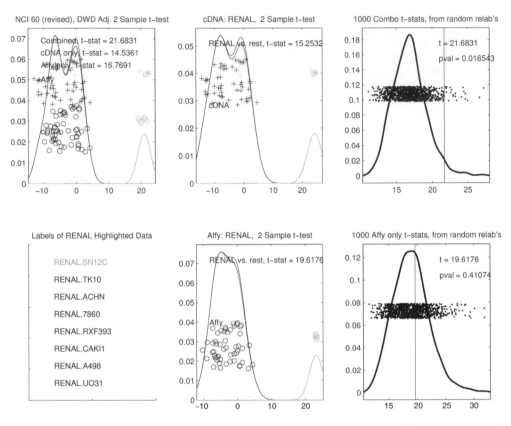

Figure 14.5 DWD and permutation based hypothesis tests of statistical significance of the renal cluster. The main lesson is that the combined data statistical inference is more powerful than for the Affymetrix only data. (Color version available as Figure S14.5 on website.)

are recomputed. However, the combined data still appear to provide more statistical power, which again seems to confirm the value of combining data.

The bottom center panel of Figure 14.5 provides the same analysis, this time for the Affymetrix only data. Again there is improvement compared to using the combined data DWD direction (top left-hand panel), but the resulting t statistic is still inferior to that for the combined data, again suggesting the combination has been worthwhile. Again the Affymetrix only t statistic is also larger than the cDNA only statistic, showing improved power for that platform.

While the t statistics give some useful information, they are still not conclusive. In particular, the sample sizes for the combined data are twice the size as for the individual platform data sets. Thus the t statistics are not comparable. This problem is overcome in the right-hand panels of Figure 14.5, using a permutation methodology to compute p-values. In particular, the data are randomly relabeled, to give subclasses of the same size as the leukemia subpopulation. The DWD direction is recomputed for these relabeled data and the corresponding t statistic is computed. This process is repeated 1000 times. The values of the t statistics are shown as dots in the right-hand panels, with dots in the top panel for the combined data, and dots in the bottom panel for the Affymetrix only data. This statistical methodology has been called DiProPerm by Wichers *et al.* (unpublished; further information is available from marron@email.unc.edu, and a Matlab routine is available at http://stat-or.unc.edu/webspace/miscellaneous/marron/Matlab7Software/BatchAdjust/DiProPermSM.m). The same plot for the cDNA data is not shown, since the cDNA t statistics were always smaller, suggesting less statistical power. Note that the numerical values of the t statistics using relabeled data are much larger (around 15–25) than is typical for the usual t distribution (up to 2 in absolute value). This is because the DWD direction seeks to strongly separate the labeled classes, so the distribution of the dots is the distribution of the t statistic under the null hypothesis of a nonsignificant cluster.

In the top panel the t statistic of 21.7, for the combined data, from the upper left-hand panel, is compared to the null population shown as dots. Note that the actual value is larger than nearly all of the simulated dots, showing that this t statistic is clearly statistically significant. The proportion of these could be used as an empirical p-value, but in other cases this gives a value of 0 too frequently. For better relative comparison, a Gaussian distribution is fitted to the population of simulated t statistics based on the combined data (dots in the upper right-hand panel), and corresponding Gaussian quantiles are used. The same method is used for computing a p-value for the Affymetrix only data in the bottom right-hand panel. Note that the combined data p-value of 0.02 is statistically significant, while the Affymetrix p-value of 0.41 is not statistically significant. This shows conclusively that much improved statistical power comes for distinguishing the renal cluster from the DWD combination of the data, relative to testing this hypothesis on the basis of either platform alone.

Similar analyses are available for each of the other seven cancer types among the NCI 60 data, from the 'DWD Across Platform Adjustment of the NCI-60 Data' link on the Detailed Graphics web page. A summary of the results appears in Table 14.2.

The results from Figure 14.5 are summarized in the second row of Table 14.2. The other rows contain similar summaries for the other cancer types. The second, third and fourth columns summarize the projected t statistics, with the largest for each cancer type

Table 14.2 Summary of cluster significance results of NCI 60 cancer types, showing that combined data almost always give improved statistical power, relative to individual platform analyses, in terms of t statistics and p-values.

Type	cDNA t	Affy. t	Comb. t	Affy. p	Comb. p
Melanoma	36.8	39.9	51.8	e-7	0
Leukemia	18.3	23.8	27.5	0.12	0.00001
NSCLC	17.3	25.1	23.5	0.18	0.02
Renal	15.6	20.1	22.0	0.54	0.04
CNS	13.4	18.6	18.9	0.62	0.21
Ovarian	11.2	20.8	17.0	0.21	0.27
Colon	10.3	17.4	16.3	0.74	0.58
Breast	13.8	19.6	19.3	0.51	0.16

being underlined. Note that the cDNA t statistics are always less than the others, again suggesting lower statistical power for all of the cancer types. But for the Affymetrix versus combined comparison, the results are very close, with each being better for three cancer types, while they are very close for the remaining two types. The p-values are shown in the final column of Table 14.2, with combined data p-values almost always being better than the single-platform Affymetrix p-values (except for ovarian cancer, where the results are very close). These different impressions given by the t statistic and by the p-value show that it is important to do the more complex permutation test, in order to fully understand the improvement of the combined data over single-platform data, and to determine cluster significance.

Note that different cancer types yield different levels of significance. The melanoma cluster, seen to be very strong in Figure 14.3, is clearly significant in both the combined and the single-platform tests. For three types (leukemia, non-small cell lung cancer, and renal cancer), the difference between combined and single-platform data is critical, with the combined data always giving the significant result. For the other four cancer types neither data set flags the cluster as statistically significant. This is not surprising for breast cancer, because it is known that there are several quite different cell types present (Ross and Perou 2001).

One might object to the fact that DWD has been used for both bias adjustment, and also lies at the core of the hypothesis test for verification of the method. But this is not an issue because in the first step the DWD direction is subtracted out, so any potential interaction between the two will be negative.

This clear success in cross-platform combination of microarray data appears to contradict the previously expressed view that this is impossible. In the next section a toy example is presented that will resolve this apparent contradiction, by showing that it is caused by the gene-by-gene methods of analysis used in previous studies not providing sufficient insights into the multivariate nature of these data sets.

14.4   Gene-by-Gene versus Multivariate Views

In this section the above apparent contradictions are resolved, and it is seen that gene-by-gene analyses need to be regarded with healthy skepticism in the analysis of microarray data, because the data are intrinsically multivariate in nature.

The simulated data set studied here has 4000 genes (dimensions), and is intended to reflect one important biological effect, but with gene expression measured across two platforms. There are 30 samples from each platform, split evenly between the two clusters, hence 15 points in each simulated biological cluster. Each sample is generated with independent Gaussian entries (simulating gene expression values), with standard deviation 1. The means of these entries are taken to be ± 0.2, in such a way that there are four clusters, where pairs correspond to platforms, and within each pair, the clusters simulate an important biological difference. Note that the very small difference in the means of the entries is an order of magnitude less than the noise level, so that it is essentially invisible to a gene-by-gene analysis. This is seen via both a gene-by-gene scatterplot view, and by a conventional heatmap, using clustering and TreeView as outlined at http://genome.med.unc.edu:8080/caBIG/DWDWebPage/DWDNCI60.htm. However, both the simulated cross-platform effect and the simulated biological effects have been designed

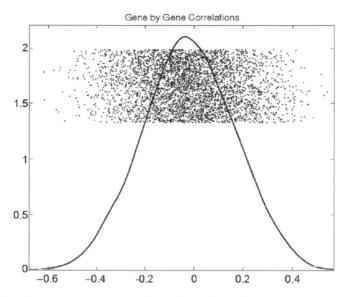

Figure 14.6   Gene-by-gene correlation analysis of simulated data, showing no significant correlations, as in an earlier analysis of the NCI 60 data. For details on the axes, see Section 14.4. (Color
version available as Figure S14.6 on website.)

to be multivariate in nature, so it is seen below that these both show up, and can be adjusted
for, using a proper multivariate view.

Figure 14.6 shows that the simulated data give the same conclusion as those of an
earlier study, when a gene-by-gene correlation analysis is done. Each dot in Figure 14.6
corresponds to one dimension (i.e. gene) of the simulated data shown in Figure 14.7.
The horizontal coordinate of the dot is the sample correlation of the simulated expression
levels of that gene, for the paired data across the platforms. The vertical coordinate is
again a random value, which provides visual separation. Most of the correlations tend to
be clustered around 0. There is some variation, but the amount of this is approximately
what would be expected at random. This is essentially a reconstruction of a gene-by-gene
analysis of these simulated data.

The limitation of the gene-by-gene correlation analysis is made clear in the PCA multivariate scatterplot view of these data (shown in Figure 14.7). These plots show various
projections on the first three PCs. Note that the first two PCs (top center panel) contain
the deliberately constructed structure in the data. In particular, the platform effect (indicated by the gray and black colors) is clear, as is the strong simulated biological effect,
shown as two clusters (indicated by the plus and circle symbols). The fact that platform
adjustment can be successful is indicated by the fact that the lines (connecting paired
samples) are approximately parallel.

After DWD adjustment, the platform effect essentially disappears (this can be seen in a
plot shown at http://genome.med.unc.edu:8080/caBIG/DWDWebPage/DWDNCI60.htm).
The four clusters visible here become two clusters. The paired data are not exactly on top
of each other, but the connecting line segments are all much smaller than the distances
between clusters. This is a simulated reconstruction of the lessons learned in Figure 14.4.

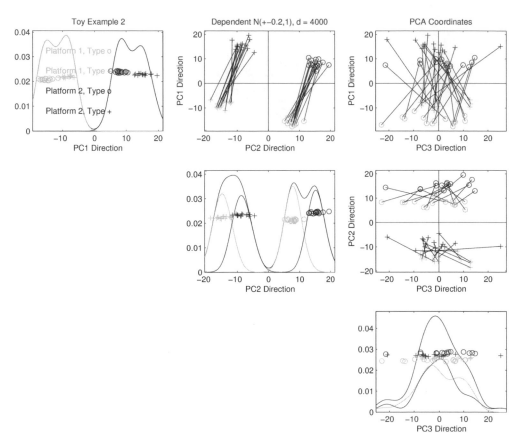

Figure 14.7   PCA scatterplot view of simulated data, showing that simulated strong platform and biological effects are present in these data. (Color version available as Figure S14.7 on website.)

Finally, we revisit the correlation analysis. The limitation of the earlier analysis was that it only looked in the gene-by-gene direction. The same suggestion of no correlation still applies for the DWD adjusted data. However, the relevant direction is not gene-by-gene, but instead in the PC 1 direction. When the correlation is computed on the data projected in this direction, the correlation becomes 0.98, meaning that there is very strong information in the data when one looks in the correct direction. This explains the above apparent paradox.

Further details of this analysis, including more detailed views, and access to the data set itself, are available from the 'DWD Across-Platform Adjustment of the NCI-60 Data' link at http://genome.med.unc.edu:8080/caBIG/DWDWebPage/DWDNCI60.htm. A minor technical note is that in the earlier study a few pairs of samples were eliminated, because there were some questions about data quality. While that was a good decision, for the point that they were trying to make, we have chosen to include all of the data. The reason is that we wish to make the point that we *can* do cross-platform normalization, and believe it is important to demonstrate this even in the presence of a few samples of questionable quality.

The ideas discussed in this section are related to the geometric representation ideas of Hall *et al.* (2005). That paper develops a mode of nonstandard asymptotic analysis that is relevant to data of high dimension and low sample size, such as microarrays. Some surprising underlying structure is shown, which is then used for classical mathematical statistical purposes, such as comparison of discrimination rules.

14.5   Conclusion

This chapter has demonstrated the utility, in the context of microarray data analysis, of DWD for all of bias adjustment, visualization, and as the basis of the DiProPerm hypothesis test. These points were made using both the NCI 60 data and simulated examples. Careful visualization demonstrates why this result is in fact consistent with earlier apparently negative results.

A Matlab version of DWD is available from the 'Matlab software for DWD adjustment' link on the UNC Adjustment of Systematic Microarray Data Biases web page, https://genome.unc.edu/pubsup/dwd/. A more portable JAVA version of DWD has been developed as part of the caBIG Project; see https://cabig.nci.nih.gov/tools/DWD. Many more related figures are available at the UNC Adjustment of Systematic Microarray Data Biases web page.

15

Toward Integration of Biological Noise: Aggregation Effect in Microarray Data Analysis

Lev Klebanov and Andreas Scherer

Abstract

Aggregation effect in microarray data analysis distorts the correlations between gene expression levels and, in some sense, plays a role of technical noise. This aspect is especially important in network and association inference analyses. However, it is possible to construct statistical estimators which take aggregation into account to generate 'clean' covariance of expression levels. Based on this estimator, we provide a method to find gene pairs having essentially different correlation structure.

15.1 Introduction

It is widely believed that a high level of technical noise in microarray data is the most critical deterrent to the successful use of this technology in studies of normal and abnormal biological processes. In particular, the notorious lack of reproducibility of lists of detected genes across platforms and laboratories, as well as validation problems associated with prognostic signatures, is frequently attributed to this 'flaw' of microarray technology (Marshall 2004; Cobb 2006). This common belief also serves as a motivation for complex normalization procedures. Strange as it may seem, it was not until recently that a specially designed metrological study was reported by the MicroArray Quality Control (MAQC) Consortium in *Nature Biotechnology* (Shi *et al.* 2006). This long overdue study, led by Food and Drug Administration scientists, provides valuable information for technical noise assessment. Klebanov analysed a subset of Affymetrix GeneChip data included in the MAQC data set and came to the conclusion that in this particular, well-controlled study the variability in microarray data caused by technical noise is low and its role in statistical methodology of data analysis, exemplified by estimation of correlation coefficients, is far from critical (Klebanov *et al.* 2007).

Typically, the goals of microarray studies are: (1) to identify genes that are differentially expressed (DE) between two or more treatments or conditions; and/or (2) to study the dependence between expression levels of different genes (e.g. for network or pathway analysis).

The solution of the first problem is usually based on the theory of statistical hypothesis testing. Since the number of genes in a microarray study is typically in the tens of thousands, a severe multiple-testing problem results. Several trends have dominated common practice in analyzing microarray data in recent years (Allison *et al.* 2006). The currently practiced methods of significance testing in microarray gene expression profiling are unstable and tend to be low in power. These undesirable properties are due to the nature of multiple-testing procedures, as well as extremely strong and long-ranged correlations between gene expression levels. In an earlier publication, Klebanov and Yakovlev (2007) identified a special structure in gene expression data that produces a sequence of weakly dependent random variables. This structure, termed the δ-sequence, lies at the heart of a new methodology for selecting differentially expressed genes in nonoverlapping gene pairs. The proposed method has two distinct advantages: it leads to dramatic gains in terms of the mean numbers of true and false discoveries, and in the stability of the results of testing; and its outcomes are entirely free of the log-additive array-specific technical noise. Klebanov *et al.* (2008) demonstrated the usefulness of this approach in conjunction with the nonparametric empirical Bayes method. The proposed modification of the empirical Bayes method leads to significant improvements in its performance. The new paradigm arising from the existence of the δ-sequence in biological data offers considerable scope for future developments in this area of methodological research.

The solution to the second problem is based on the study of correlations between gene expression levels, as well as on gene co-expressions. Inferring gene regulatory networks from microarray data has become a popular activity in recent years, resulting in an ever-increasing volume of publications. There are many pitfalls in network analysis remaining either unnoticed or scantily understood. One of the problems we see is that inferences about 'microscopic' processes of transcription within individual cells are made from 'macroscopic' observations by gene expression measurements, and hence a 'mixing effect' derived from signal summation should be considered confounding. Another problem we see is within the issue of general technical noise, but also multiple targeting. A critical discussion of such pitfalls is long overdue.

The same caveat is of even greater concern with reference to more sophisticated methodologies that are designed to extract more information from the joint distributions of expression signals, Bayesian network inference being a relevant example. Chu *et al.* (2003) pointed out the important fact that the measurements of mRNA abundance produced by microarray technology represent aggregated expression signals and, as such, may not adequately reflect the molecular events occurring within individual cells. To illustrate this conjecture, the authors proceeded from the observation that each gene expression measurement produced by a microarray is the sum of the expression levels over many cells. In fact, on each array we observe a sum of gene expressions coming from different cells. The number of cells which is used as starting material is not controlled and therefore is unknown. This circumstance gives the so-called aggregation effect – each expression level is a sum of a random number (number of cells used per array) of random expression levels, coming from a fixed gene. This random number of cells plays a role similar to that of technical noise (Klebanov and Yakovlev 2008). However, there is another element in the 'noise' problem. The gene expression measurements per array are produced by cells being in different stages of the cell cycle. Therefore, we observe a mixture of many different states, which changes the correlation structure between different genes. The aggregation effect may be now considered as consisting of two parts. The first part of the effect is generated by a randomness of the number of cells on an array. This part should be considered noise. To avoid corresponding correlation changes one has to fix the number of cells on each array. The second part of the effect is explained by the heterogeneity of the cell population (cells are in different stages of the cell cycle). This part may be considered as a nuisance in the case where one would like to study the dependency between genes in a cell. However, it may be very informative if one is studying a dependency between cells.

Klebanov and Yakovlev (2008) provided a theoretical consideration of the random effect of signal aggregation and its implications for correlation analysis and network inference. The authors attempted to assess quantitatively the magnitude of this signal aggregation from real data. Some preliminary ideas were offered to mitigate the consequences of random signal aggregation in the analysis of gene expression data. Resulting from the summation of expression intensities over a random number of individual cells, the observed signals may not adequately reflect the true dependence structure of intra-cellular gene expression levels needed as a source of information for network reconstruction.

In this chapter we discuss feature aggregation in microarray data, which investigators need to be aware of when embarking on a study of putative associations between elements of networks and pathways. We believe that the present discussion pinpoints the crux of the difficulty in correlation analysis of microarray data and network inference based on correlation measures.

15.2   Aggregated Expression Intensities

Let v be the number of cells contributing to the observed expression signal U and denote by X_i the expression level of a given gene in the ith cell. The notation Y_i is used for the second gene in a given pair of genes. A simplistic model of the observed expression signals in this pair is given by

$$U = X_1 + \cdots + X_v, \quad V = Y_1 + \cdots + Y_v, \tag{15.1}$$

where X_i and Y_i are two sequences of independent and identically distributed (i.i.d.) random variables, while X_i and Y_i in each pair (X_i, Y_i) may be dependent with joint distribution function $F(x, y)$. It is natural to suppose that X_i and Y_j are independent for $i \neq j$. Limiting themselves to the case where v is nonrandom, Chu *et al.* (2003) showed that, except for some very special and biologically irrelevant cases, the Markov factorization admitted by the expression levels within individual cells does not survive the summation (aggregation), thereby stymieing any network inference based on the joint distribution. The importance of this observation cannot be emphasized enough. Obtained in Klebanov and Yakovlev (2008), the formula gives an expression for the correlation coefficient $\rho(U, V)$ between U and V:

$$\rho(U, V) = \frac{E\nu \text{Cov}(X, Y) + \sigma_\nu^2 EX\, EY}{\sqrt{E\nu\sigma_X^2 + \sigma_\nu^2(EX)^2}\sqrt{E\nu\sigma_Y^2 + \sigma_\nu^2(EY)^2}} \tag{15.2}$$

where E is used for the expected value and $\text{Cov}(X, Y)$ is the covariance between X and Y (i.e. between each of the pairs $(X_i, Y_i), i = 1, \ldots, \nu$). In Klebanov and Yakovlev (2008) it was shown that $\rho(U, V) = \rho(X, Y)$ if and only if $\sigma_\nu^2 = 0$.

If the hybridization reaction reaches equilibrium, the random variable ν can be interpreted as the total number of cells from which the total RNA is extracted. In the practical use of microarray technology, however, the reaction is typically stopped before equilibrium has been reached. Therefore, the random variable ν is unobservable.

One can see from equation (15.2) that the correlations between observable U and V are essentially different from those between X and Y, especially because the random variable ν is unobservable, and one cannot estimate $E\nu$ and σ_ν^2. This fact shows that the aggregation effect plays a role of a noise; it does not allow one to reconstruct both individual expression levels of different genes as well as the true correlations between gene expression levels.

However, Klebanov and Yakovlev (2008) have shown that the correlations between U/ν and V/ν coincide with those between X and Y, which gives us a possibility to recover true levels of the rations of gene expressions, and explains why the methods based on δ-sequences are so effective.

Unfortunately, we cannot recover the correlations between X and Y. Our goal in this paper is to show what type of information on the connection between expression levels it is possible to obtain from aggregated data.

15.3   Covariance between Log-Expressions

Here we will change our notation slightly. Denote by $X_j(k)$ the expression of kth gene in the jth cell. Thus, $U(k) = \sum_{j=1}^{\nu} X_j(k)$ is the total expression of kth gene obtained from all cells on the array. Taking logarithms to base 2 (and denoting this simply as log), we obtain

$$S(k) = \log U(k) = Z(k) + \log \nu,$$

where

$$Z(k) = \log \frac{\sum_{j=1}^{\nu} X_j(k)}{\nu}.$$

The quantities $\frac{\sum_{j=1}^{\nu} X_j(k)}{\nu}$ are of special interest to ud us because the correlations between them for different values of k are the same as between corresponding $X(k)$ (see Klebanov and Yakovlev 2008).

For an arbitrary k, s we have

$$\text{Var}(S(k) + S(s)) = \sigma_{Z(k)}^2 + \sigma_{Z(s)}^2 + 2\text{Cov}(Z(k), Z(s)) + 4\sigma_\nu^2,$$
$$\text{Var}(S(k) - S(s)) = \sigma_{Z(k)}^2 + \sigma_{Z(s)}^2 - 2\text{Cov}(Z(k), Z(s)).$$

Denoting

$$Q(k, s) = \frac{1}{4} \left(\text{Var}(S(k) + S(s)) - \text{Var}(S(k) - S(s)) \right),$$

we obtain

$$Q(k_2, s_2) - Q(k_1, s_1) = \text{Cov}(Z(k_2), Z(s_2)) - \text{Cov}(Z(k_1), Z(s_1)) \qquad (15.3)$$

for all k_1, k_2.

Equation (15.3) allows us to exclude the influence of the random aggregation effect from the difference in covariance between different pairs of gene expression levels. If we fix k_1 and s_1, it is possible to see how large is the region of changes in covariance for other pairs of gene expression levels. Below we fix $k_1 = 1$, $s_1 = 2$. Equation (15.3) gives one a possibility to avoid the aggregation effect while estimating the covariance increments.

The covariance between expression levels of different gene pairs was calculated for two sets of previously published microarray data. Yeoh *et al.* (2002) investigated microarray data (Affymetrix HG-U95Av2) from bone marrow cells of pediatric patients with different types of acute lymphoblastic leukemia (ALL). Their data set consisted of 327 diagnostic bone marrow samples from a variety of ALL subtypes. Yeoh *et al.* addressed the question whether gene expression profiling of these samples could assist the tedious work of ALL-subtype classification which is done by immunophenotyping, cytogenetics and other molecular diagnostics procedures. Correct stratification of the patients is essential for successful treatment adjustment, as patients with different ALL subtypes react differently to chemotherapy, and hence the treatment needs to be personalized. In the following, we present data we obtained by using the data from patients with TEL-AML1 (79 cases) and with hyperdiploidy (64 cases). The TEL-AML1 phenotype is a B lineage leukemia which contains a t(12;21)(p12;q21) chromosomal rearrangement creating a chimeric protein; the hyperdiploidy phenotype is a hyperdiploid karyotype with more than 50 chromosomes.

Figure 15.1(a) shows mean differences between covariance for probe set pairs (k, s) and pair (1,2), that is,

$$\frac{1}{m-2} \sum_{k=3}^{m} (Q(k, s) - Q(1, 2)),$$

where m was chosen to be 2000, for the hyperdiploidy data set ($s = 1, \ldots, m$). On the x-axis is the probe set number, and on the y-axis the value of mean covariance difference (this also applies to the rest of Figure 15.1). We see that the mean differences are mostly positive, and tend to be close to 0.08. The mean value is 0.077, with standard deviation 0.017.

Figure 15.1(b) shows mean differences between covariance for probe set pairs (k, s) and pair (1,2), calculated in the same way as for Figure 15.1(a), for the TEL/AML-1 data set ($s = 1, \ldots, m$). In this case the mean differences are essentially smaller than for the hyperdiploidy case. They remain mostly positive, but tend to be close to 0.05. The mean value is 0.044, with standard deviation 0.014. The structure of observed mean values, given in Figures 15.1(a) and 15.1(b), shows that the correlations between gene expression levels are mostly positive, and this is not a 'side effect' of the aggregation, because the sign of covariance is the same as that of correlation.

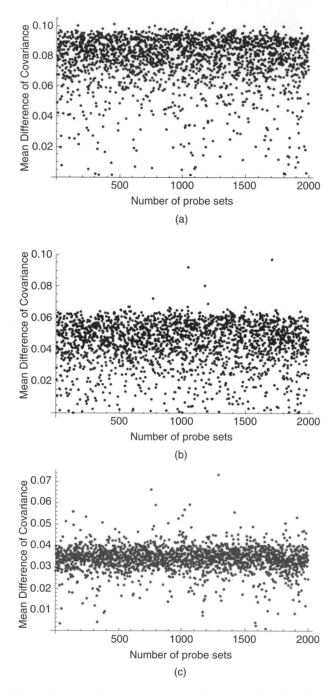

Figure 15.1 (a) Mean differences between covariance for gene pairs of probe sets (k, s) and pair (1,2) for the hyperdiploidy data set. (b) Mean differences between covariance for gene pairs (k, s) and pair (1,2) for TEL/AML1 data set. (c) Mean differences between covariance of TEL/AML1 and hyperdiploidy expression levels.

Hu *et al.* (2009) tried to apply correlation vectors (i.e. the vectors whose coordinates are the correlation of expression levels of a given gene with all expression levels of all other genes) to find differentially correlated genes in two states. Unfortunately, the method does not take into account the aggregation effect described above. One can partially correct the situation by using mean covariance differences instead of the correlation vector norms, or just covariance instead of correlation. Figure 15.1(c) shows the mean differences $\frac{1}{m-2}\sum_{k=3}^{m}(Q(k, s) - Q(1, 2))$ for hyperdiploidy and the same characteristic for TEL/AML1 databases. We see the mean values are concentrated near 0.035 (the precise mean value of all differences is 0.034, with standard deviation 0.0059). Visually, the mean differences between expression level covariances are much more closely concentrated near 0.034 than differences inside each separate database.

This follows also from the fact that the standard deviation in the case of comparing different databases is the smallest. It allows us to use these mean values to find large changes in the covariance structure of gene expression levels. We may select such genes for which the changes in mean covariance are close to maximal. To see what changes are large enough, it is possible to use a cross-validation procedure. The corresponding selection procedure works much more quickly than used by Hu *et al.* (2009).

The main idea for the procedure proposed above is that the correlation–covariance structure of unchanged genes remains the same, but it has to be different for some differentially expressed genes. A question arises whether the procedure described will be stable with respect to changes of gene pair to compare with. For the construction of Figure 15.1 we used the pair (1,2). The situation remains almost unchanged if we use other 'ordinary' pairs instead. By 'ordinary' pairs mean those pairs for which corresponding points on Figure 15.1(c) are close to horizontal line $y = 0.034$. The only change is that instead of 0.034 we will have another number.

We should mention that we also analysed data from the T-ALL phenotype in Yeoh's data set, as well as other data sets, and obtained results similar to the ones presented here.

15.4   Conclusion

We have provided theoretical considerations on a fundamental issue in statistics, called Simpson's paradox, which describes the reversal of inequalities caused by fractionation of data and study groups, and we apply these to microarray technology (Simpson 1951). Signal aggregation is a phenomenon which has so far attracted little attention in the microarray literature. Whether the effect is extreme or not, the important point is to recognize and incorporate such signal source for proper inference. The usefulness of inference on genetic regulatory structures from microarray data depends critically on the ability of investigators to overcome this obstacle in a scientifically sound way. The method we propose is closely connected to an approach by Hu *et al.* (2009), which is based on distribution of correlation vectors.

While we describe theoretical implications of aggregation effect for network inference, we are aware that in practice the caveats of signal aggregation cannot yet be tackled. The problem with hybridization-based technologies is that the latent parameter ν is not accessible to measurement.

The future of inferential network analysis hinges on the ability to surmount the obstacle described in this chapter either by means of mathematics and/or through radical technological improvements.

Acknowledgements

This research is supported by grant MSM 002160839 from the Ministry of Education, Czech Republic.

16

Potential Sources of Spurious Associations and Batch Effects in Genome-Wide Association Studies

Huixiao Hong, Leming Shi, James C Fuscoe, Federico Goodsaid, Donna Mendrick, and Weida Tong

Abstract

Genome-wide association studies (GWAS) use dense maps of single nucleotide polymorphisms (SNPs) that cover the entire human genome to search for genetic markers with different allele frequencies between cases and controls. Given the complexity of GWAS, it is not surprising that only a small portion of associated SNPs in the initial GWAS results were successfully replicated in the same populations. Each of the steps in a GWAS has the potential to generate spurious associations. In addition, there are batch effects in the genotyping experiments and in genotype calling that can cause both Type I and Type II errors. Decreasing or eliminating the various sources of spurious associations and batch effects is vital for reliably translating GWAS findings to clinical practice and personalized medicine. Here we review and discuss the variety of sources of spurious associations and batch effects in GWAS and provide possible solutions to the problems.

16.1 Introduction

The aim of GWAS is to identify genetic variants, usually single nucleotide polymorphisms (SNPs), across the entire human genome that are associated with phenotypic traits, such as disease status and drug response. A statistically significant difference in allele frequencies of a genetic marker between cases and controls implies that the corresponding region of the human genome contains functional DNA sequence variants that make contributions to the phenotypic traits of interest. The International HapMap project determined genotypes

Batch Effects and Noise in Microarray Experiments: Sources and Solutions edited by A. Scherer
© 2009 John Wiley & Sons, Ltd

of over 3.1 million common SNPs in human populations and computationally assembled them into a genome-wide map of SNP-tagged haplotypes (International HapMap Consortium et al. 2007). Concurrently, high-throughput SNP genotyping technology advanced to enable simultaneous genotyping of hundreds of thousands of SNPs. These advances combine to make GWAS a feasible and promising research field.

GWAS have recently been applied to identify common genetic variants associated with a variety of phenotypes (Frayling et al. 2007; Saxena et al. 2007; Zeggini et al. 2007; Scott et al. 2007; Sladek et al. 2007; Steinthorsdottir et al. 2007; Wellcome Trust Case Control Consortium et al. 2007; Mullighan et al. 2007; Buch et al. 2007; Gold et al. 2008). These findings are valuable for scientists in elucidating the allelic architecture of complex traits in general. However, replication of GWAS showed that only a small portion of association SNPs in the initial GWAS results can be replicated in the same populations. For example, in replication studies of GWAS for type 2 diabetes mellitus, only 10 out of the 77 association SNPs tested (Zeggini et al. 2007), 10 out of the 80 association SNPs tested (Scott et al. 2007), eight out of the 57 association SNPs tested (Sladek et al. 2007), and two out of the 47 association SNPs tested (Steinthorsdottir et al. 2007) were successfully replicated. Obviously, there are many false positive associations in GWAS results. Moreover, the lists of positive associations identified in different GWAS for a same disease, such as type 2 diabetes mellitus, were quite different. Thus, there are potentially many false positive associations found (Type I errors) and true positive associations missed (Type II errors) in GWAS results, limiting the potential for early application to personalized medicine.

GWAS are complicated processes depicted by a simplified cartoon shown in Figure 16.1. Each step has the potential to generate spurious associations.

GWAS are based on the common trait/common variant hypothesis which implies that the genetic architecture of complex traits consists of a limited number of common alleles, each conferring a small increase in risk to the individual (Reich and Lander 2001). Therefore, the likelihood of detecting an association is usually small and thus requires a large sample size to achieve adequate statistical power. However, the current technology cannot genotype all samples in an association study at once and thus samples are usually partitioned and genotyped at different times (batches). In addition, the raw data files (CEL files) are very large (hundreds of gigabytes) and must be partitioned into batches containing subsets of the entire data set for genotype calling. Variation in both these processes caused by the need to partition the samples for experimental analysis as well as for statistical analysis has the potential to induce Type I and Type II errors in GWAS.

This chapter reviews each major step in the GWAS workflow (Figure 16.1) that has the potential to generate spurious associations. Examples of potential sources of spurious associations are discussed in detail.

16.2   Potential Sources of Spurious Associations

Given the complexity of GWAS, multiple sources of Type I and Type II errors can arise in their results (associated SNPs). Table 16.1 lists potential sources of errors that can lead to spurious associations, as well as possible batch effects. In this section, the potential sources and possible solutions of spurious associations in GWAS are reviewed and discussed.

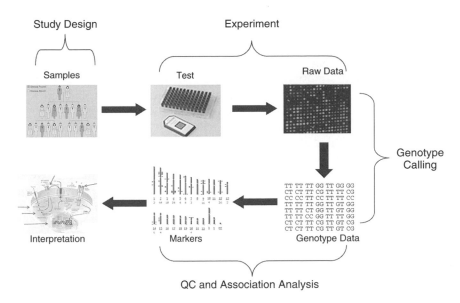

Figure 16.1   Overview of workflow for a genome-wide association study. First, cases and controls are selected. Then the samples from cases and controls are subjected to genotyping. The raw data from the genotyping experiment are used to determine genotypes for SNPs of the samples. Downstream association analyses are conducted using the genotype data to identify significantly associated SNPs for the trait in study. Finally, interpretation of the possible biological functions of significant SNPs may aid understanding of the trait.

Table 16.1   Potential batch effects and sources of spurious associations in GWAS.

Step in GWAS	Potential sources
Study design	Sample size Selection and ascertainment of participants Population substructure
Experiment	Variation in genotyping platforms Variation in generations of a genotyping platform Reproducibility between laboratories Technical repeatability of genotyping Genotyping batch effect
Genotype calling	Variation in calling algorithms Consistency between versions of a calling algorithm Variation in parameter settings of a calling algorithm Genotype calling batch effect Different statistical tests Different genetic tests

16.2.1   Spurious Associations Related to Study Design

16.2.1.1   Sample Size

Although there is increasing interest in the application of GWAS methodologies to population based cohorts, most of the first wave of GWAS used case–control study designs, in which allele frequencies in patients with a disease were compared with those in a comparison healthy group. The blooming of GWAS, witnessed by a substantial crop of GWAS publications in prestigious journals such as *Science* and *Nature* in recent years, has demonstrated that the effects resulting from common SNP associations in GWAS are usually small. Therefore, sample sizes in GWAS are large, usually at the level of thousands, to achieve adequate statistical power. A small sample size reduces the statistical power, which may mask true associations. Although an exact consensus on sample size has not been reached, it is clear that the larger the sample size, the better, all other things being equal.

16.2.1.2   Selection Bias

Improper selection of participants for GWAS is a potential source of obfuscation and variability in GWAS design. Misclassifying participants into case and control groups can markedly reduce study power and result in spurious associations, particularly when a large number of nondiseased individuals are misclassified as diseased (Pearson and Manolio 2008). When selecting cases, careful participant ascertainment is necessary, especially for diseases that are difficult to diagnose reliably. However, a bias in selection of cases may arise if individuals are selected to over-minimize phenotypic heterogeneity by focusing on extreme and/or familial cases, resulting in decreasing statistical power (Howson *et al.* 2005; Li *et al.* 2006). The results of such studies may not be applicable to other populations. Controls should be selected from the same population as affected individuals. Optimal selection of controls remains controversial, although the accumulating empirical data indicate that many commonly expressed concerns have been overstated (McCarthy *et al.* 2008). Effectiveness of non-population-based controls was demonstrated by a study in which the 3000 UK controls (as 'common control') were compared with 2000 cases with each of seven different diseases (Wellcome Trust Case Control Consortium 2007). This suggests that initial identification of association with disease may be resistant to these biases of using non-population controls, especially given subsequent successful replication of some associations in studies using more traditional control groups (Scott *et al.* 2007; Hakonarson *et al.* 2007; Samani *et al.* 2007). Of more concern in the improper selection of cases and controls may be nongenetic covariates – such as smoking (Dewan *et al.* 2006) and obesity (Frayling *et al.* 2007) – which may be confounded with the outcome and subsequently generate false positive associations.

16.2.1.3   Stratification of Cases

Improper population stratification of cases and controls is an additional potential source of spurious associations related to GWAS design. Population stratification (imbalances in populations in cases relative to controls) inflates the Type I error rate around variants that are informative about the population substructure (Price *et al.* 2006), but its influence is a matter of debate (Thomas and Witte 2002; Wacholder *et al.* 2002). Several statistical tools have been developed to correct for population stratification (Cardon and Palmer 2003; Price *et al.* 2006; Zheng *et al.* 2006) which are now incorporated into rigorous GWAS analyses.

16.2.2 Spurious Associations Caused in Genotyping Experiments

A fundamental component of GWAS is the genotyping experiment (Figure 16.1) that determines genotypes of SNPs in human subjects. Each stage in a genotyping experiment is fraught with the potential for error and bias. Genotyping errors generated in this step, especially if distributed differentially between cases and controls, are potential sources of spurious associations in GWAS findings and have to be diligently sought and corrected.

The debates about SNP selection (Barrett and Cardon 2006; Pe'er *et al.* 2006) that dominated the early discussions on GWAS have cooled down. The issue has boiled down to the choice of genotyping platforms, mainly between two leaders (Affymetrix and Illumina), for conducting genotyping experiments. Differences in SNPs coverage between Affymetrix and Illumina are a potential source of Type II errors (losing significant associations). Since the designs of probe sets on SNP arrays, as well as the experimental protocols, are quite different between these two platforms, the consistency of common SNPs interrogated in both platforms is vital for obtaining reliable GWAS results. Inconsistency of this type of SNPs may cause both Type I and Type II errors. Solid evaluation of variations between platforms has not been reported and is expected to be thoroughly assessed in the future.

Different generations of genotyping platforms from both Affymetrix and Illumina have been delivered to the market and used in GWAS. For example, the Affymetrix GeneChip Human Mapping 500K array set (Affy500K) has been used in published GWAS (Saxena *et al.* 2007; Wellcome Trust Case Control Consortium 2007; Mullighan *et al.* 2007; Buch *et al.* 2007), but recently Affymetrix released the Genome-Wide Human SNP 6.0 array (Affy6) to the market. Since it is important to know whether genotypes determined with these two generations of SNP array are consistent, solid evaluation of their consistency is expected to be conducted in the future.

Reproducibility of genotyping across laboratories is another concern. Realizing the high cost of genotyping at this time, none of the current GWAS results have been confirmed with the same samples and genotyping technologies by other laboratories. Although some of the GWAS findings were successfully reproduced with different samples from the same population and different genotyping technology, the reproducibility of genotyping across laboratories remains unclear and needs be assessed in the future.

16.2.3 Spurious Associations Caused by Genotype Calling Errors

A genotype calling algorithm assigns genotypes for SNPs from raw intensity data. Many algorithms have been developed. One of the fundamental questions is how consistent the different calling algorithms are, though each of the algorithms reported a high call rate and accuracy. We evaluated the concordance of genotype calls from three algorithms from Affymetrix (DM, BRLMM, and Birdseed; see details at http://www.affymetrix.com/

products services/software/specific/genotyping console software.affx) that were released along with its recent three generations of SNP arrays, and assessed potential spurious associations caused by genotyping calling algorithms. Discordance in genotypes was found in both missing and successful calls. Our observations suggest that there is room for improvements in both call rates and accuracy of genotype calling algorithms. Different versions of the same calling algorithm have been developed and are in use for GWAS. For example, Birdseed is a model-based clustering algorithm that converts continuous intensity data to discrete genotype data. It has two different versions. Version 1 fits a Gaussian mixture model into a two-dimensional space using SNP-specific models as starting points to start the expectation–maximization (EM) algorithm, while version 2 uses SNP-specific models in a pseudo-Bayesian fashion, limiting the possibility of arriving at a genotype clustering that is very different from the supplied models. Therefore, inconsistent genotypes may be called from the same raw intensity data using different versions of the algorithm, especially when intensity data do not fit the SNP-specific model perfectly. The inconsistency in genotypes caused by different versions of a calling algorithm has the potential to cause Type I and Type II errors. Thorough evaluation of inconsistency between different versions of genotype calling algorithms and its effect on significantly associated SNPs has not yet been reported.

Genotype calling is a complicated process that usually requires many user-specified parameters to be adjusted. For example, it should be decided (a) whether normalization is conducted before a calling algorithm is applied to determine genotypes based on the intensity data of probe sets, and (b) which normalization method, if required, should be used. There are also many algorithm-specific parameters that need to be set. BLRMM, for example, first derives an initial guess for each SNP's genotype using the DM algorithm (Di *et al.* 2005) and then analyzes across SNPs to identify cases of nonmonomorphisms. This subset of nonmonomorphism SNPs is then used to estimate a prior distribution on cluster centers and variance–covariance matrices. This subset of SNP genotypes is revisited and the clusters and variances of the initial genotype guesses are combined with the prior distribution information of the SNPs in an ad hoc Bayesian procedure to derive a posterior estimate of cluster centers and variances. Genotypes of SNPs are called according to their Mahalanobis distances from the three cluster centers and confidence scores are assigned to the calls. Therefore, the parameters with which to specify the p-value cutoff for DM algorithm to seed clusters (default is set to 0.17) and the number of probe sets to be used for determining prior distribution (default is set to 10 000) influence the prior distribution on cluster centers and variance–covariance matrices. Different parameter settings will cause inconsistent genotype calls. Thus, variation in parameter settings for an algorithm is a potential source of spurious associations in GWAS and needs to be systematically assessed in the future.

16.3 Batch Effects

16.3.1 Batch Effect in Genotyping Experiment

GWAS are based on the common trait/common variant hypothesis that the genetic architecture of complex traits consists of a limited number of common alleles, each conferring a small increase in risk to the individual (Reich and Lander 2001). Therefore, the odds for an association SNP are usually very small and require a large sample size to achieve enough statistical power to reveal the true associations. GWAS usually have thousands of samples. For technical reasons, genotyping of such large sample sizes cannot be done all at once; instead, subsets of the samples are grouped into batches and genotyped at multiple times. Using different batches is fraught with the potential for error and bias in the raw intensity data which, in turn, influence the genotypes determined and the significantly associated SNPs identified. For example, in GWAS of the Wellcome Trust Case Control Consortium, samples were placed in 96-well plates and were shipped on different dates to Affymetrix Services Lab for genotyping. It was observed that the missing genotype call rates per sample on plates shipped at different dates are quite different. This difference might be due to shipping issues (e.g. temperature extremes). However, a batch effect in genotyping may be an additional source of this difference. Because batch effects in genotyping experiments have not been thoroughly evaluated, it remains uncertain whether, and at what level, the variation in batches of samples in genotyping experiments using current genotyping technologies has effects, both on raw intensity data and on GWAS findings. Therefore, efforts to assess and prevent batch-related artifacts in genotyping seem to be a high priority in GWAS, despite continuing improvements in performance of genotyping technologies.

16.3.2   Batch Effect in Genotype Calling

A key step in GWAS is to determine genotypes of SNPs using a calling algorithm. High call rates and accuracy of genotype calling are important and essential issues for the success of GWAS, since errors introduced in the genotypes by calling algorithms can inflate both Type I and Type II error rates. GWAS usually involve analyses of thousands of samples that generate thousands of raw data (CEL) files. The raw data file for one sample is tens of megabytes in size. Computer memory (RAM) limits make it unfeasible to analyze all CEL files in an association study in a single batch on a single computer. The samples are, therefore, divided into many batches for genotype calling. The variation in ways to divide samples into different batches for genotype calling has the potential to generate disparities in called genotypes that in turn cause spurious associations in GWAS results. The effects on genotype calls caused by changing the number and specific combinations of CEL files in batches and propagation of the effects to the downstream association analysis have been assessed using 270 HapMap samples and the calling algorithm BRLMM (Hong *et al.* 2008). The data can be obtained from http://www.hapmap.org/downloads/raw_data/affy500k/.

Three experiments were designed and conducted to assess the effect of batch size. In the first experiment (BS1), the 270 HapMap samples were divided into three batches based on their population groups: 90 Europeans, 90 Asians, and 90 Africans. The genotypes were called separately by BRLMM using the default parameter setting suggested by Affymetrix (CEL files from genomic DNA which had been digested with the restriction enzyme NspI and CEL files from genomic DNA which had been cut with the restriction enzyme StyI were analyzed separately). The second experiment (BS2) used a batch size of 45 samples. Genotypes were called from the CEL files from 90 European samples in two batches, each with 45 CEL files using BRLMM with the same parameter settings as in BS1. The

Table 16.2 Concordance of calls between batch sizes: successful calls for both SNP genotypes successfully called in both of the compared experiments; concordant calls (All), same genotype called in both of the compared experiments; concordant calls (Hom), homozygous genotype called in both of the compared experiments; concordant calls (Het), heterozygous genotype called in both of the compared experiments.

Comparison		BS1 vs BS2	BS1 vs BS3	BS2 vs BS3
Successful calls for both	SNPs	134258764	134187584	134265847
	%	99.338	99.285	99.343
Concordant calls (All)	SNPs	134248899	134187584	134253973
	%	99.993	99.986	99.991
Concordant calls (Hom)	SNPs	98179772	98136394	98204063
	%	99.997	99.993	99.995
Concordant calls (Het)	SNPs	36069127	36031744	36049910
	%	99.981	99.964	99.980

procedure was repeated for the Asian and African samples. In the third experiment (BS3), the batch size was 30 samples from each population groups. To evaluate the batch size effect on genotypes called, concordance of genotypes called between experiments with different batch sizes is listed in Table 16.2. It can be seen that batch size affected genotype calls and that heterozygous genotype concordances were more affected than homozygous genotype concordances.

Given that batch size affects genotypes called, the selection of samples (CEL files) to be placed in each batch can also be anticipated to alter genotype calling. The term 'batch composition effect' is used here to denote the selected CEL files within batches. BRLMM was used with default parameter settings and the CEL files of 270 HapMap samples to test batch composition effects. In the first experiment (BC1), the 270 samples were placed in three batches. One batch contained 90 samples from the same population group, Europeans, Asians, or Africans. In the second experiment (BC2), the 90 samples in each of the three population groups were evenly divided into two subgroups, with each subgroup having 45 unique samples. Genotype calling was then conducted in three batches with composition of: (i) subgroup 1 of Europeans + subgroup 1 of Asians; (ii) subgroup 2 of Europeans + subgroup 1 of Africans; and (iii) subgroup 2 of Africans + subgroup 2 of Asians. In the third experiment (BC3), the 90 samples in each of the three population groups were evenly divided into three subgroups, with each subgroup having 30 unique samples. Genotype calling was then conducted in three batches with composition of: (i) subgroup 1 of Europeans + subgroup 1 of Asians + subgroup 1 of Africans; (ii) subgroup 2 of Europeans + subgroup 2 of Asians + subgroup 2 of Africans; and (iii) subgroup 3 of Europeans + subgroup 3 of Asians + subgroup 3 of Africans. To evaluate batch composition effect on genotypes called, concordance of genotypes called between experiments with different batch compositions is listed in Table 16.3. Batch composition not only affected the genotypes called but also was more pronounced for heterozygous genotypes compared with homozygous genotypes, since the concordance for heterozygous genotypes was lower than the corresponding concordance for homozygous genotypes.

Table 16.3 Concordance of calls between batch compositions. Successful calls for both genotype successfully called in both of the compared experiments; Concordant calls (All) same genotype called in both of the compared experiments; Concordant calls (Hom) homozygous genotype called in both of the compared experiments; Concordant calls (Het) heterozygous genotype called in both of the compared experiments.

Comparison		BC1 vs BC2	BC1 vs BC3	BC2 vs BC3
Successful calls for both	SNPs	134128046	134063768	134107787
	%	99.241	99.194	99.226
Concordant calls (All)	SNPs	134109060	134036623	134095792
	%	99.986	99.980	99.991
Concordant calls (Hom)	SNPs	98050788	97992008	98016851
	%	99.989	99.983	99.993
Concordant calls (Het)	SNPs	36058272	36044165	36078941
	%	99.977	99.970	99.985

The objective of GWAS is to identify phenotype-associated genetic markers. It is critical to assess whether and how the batch effect in genotype calling propagates to the significant SNPs identified in the downstream association analysis. Three case–control mimic association analyses were conducted for each of the calling results with different batch sizes and compositions to assess propagation of batch effect in genotype calling to the associated SNPs. After removal of low-quality SNPs by quality assurance/quality control, each of the three population groups (European, Asian, and African) was set as 'case' while the other two groups were set as 'control'. Association analyses were conducted to identify SNPs that can differentiate the 'case' group from the 'control' group. Different lists of SNPs associated with a same population group, identified from chi-squared tests by two-degrees-of-freedom genotypic association using the genotype calling results with different batch sizes and compositions, were compared using Venn diagrams. The comparisons of associated SNPs obtained from calling results with different batch sizes are given in Figure 16.2. It is clear that the batch size effect on genotype calling propagated into the downstream association analyses. Moreover, it was observed that the larger the differences in two batch sizes, the fewer associated SNPs were shared by the two batch sizes.

Figure 16.3 compares the lists of significantly associated SNPs obtained using the genotypes called by the three batch compositions. The Venn diagrams demonstrate that for the same 'case–control' framework, different lists of significantly associated SNPs were identified using the genotype calling results from different batch compositions. Therefore, the batch composition effect on genotype calling propagated to the significantly associated SNPs. Moreover, it was observed that the larger the difference of genetic homogeneity between two batch compositions, the fewer significantly associated SNPs were shared by the two batch compositions.

In summary, batch size and composition affect the genotype calling results in GWAS using BRLMM. The larger the differences in batch sizes, the larger the effect. The more homogenous the samples in the batches, the more consistent the genotype calls. The inconsistency propagates to the lists of significantly associated SNPs identified in downstream association analysis. Thus, uniform and large batch sizes should be used to make

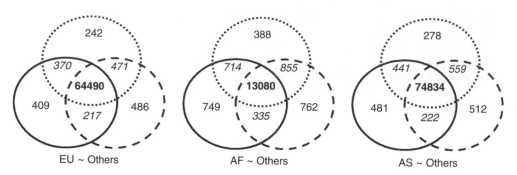

Figure 16.2 Venn diagrams for comparisons of the significantly associated SNPs identified using the genotype calling results with different calling batch sizes. The numbers in circles are the significantly associated SNPs identified in association analyses using calling results from different batch sizes solid-line circles for BS1, dotted-line circles for BS2, and dash-line circles for BS3. Numbers in bold font represent the associated SNPs shared by all three batch sizes, numbers in italic font represent the associated SNPs shared only by two batch sizes, and the numbers in normal font are the associated SNPs identified only by the corresponding batch sizes. EU~Others the association analyses results for European versus others; AF~Others the association analyses results for African versus others; AS~Others the association analyses results for Asian versus others. BS1, BS2, and BS3 are defined in the text.

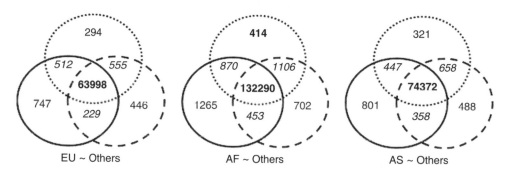

Figure 16.3 Venn diagrams for comparisons of the significantly associated SNPs identified using the genotype calling results with different calling batch compositions. The numbers in circles are the significantly associated SNPs identified in association analyses using calling results from different batch compositions solid-line circles for BC1, dotted-line circles for BC2, and dash-line circles for BC3. Numbers in bold font represent the associated SNPs shared by all three batch compositions, numbers in italic font represent the associated SNPs shared only by two batch compositions, and the numbers in normal font are the associated SNPs identified only by the corresponding batch compositions. EU~Others the association analyses results for European versus others; AF~Others the association analyses results for African versus others; AS~Others the association analyses results for Asian versus others. BC1, BC2, and BC3 are defined in the text.

genotype calls for GWAS. In addition, samples of high homogeneity should be placed into the same batch.

16.4   Conclusion

GWAS methodology is revolutionary in that it allows interrogation of the entire genome at high resolution. GWAS are not restricted by prior hypotheses regarding genetic associations with disease (Hirschhorn and Daly 2005) and have been proven suitable for identifying common SNPs with modest to large effects on phenotypic traits. Careful implementation of GWAS has identified a number of new susceptibility loci, shedding light on the fundamental mechanisms of disease predisposition. It appears that GWAS might elucidate molecular mechanisms of poorly understood common diseases. GWAS also show significant potential to redefine disease classification, which is potentially useful in personalized medicine.

However, despite the current excitement, GWAS have only ascertained a small portion of the expected genetic variance in complex traits (Altshuler and Daly 2007). Furthermore, replication studies of GWAS have shown that only a small portion of association SNPs in the initial GWAS results can be replicated in the same populations, indicating there is much left to understand about the generation and analysis of GWAS data. GWAS are complicated processes with each step filled with potential sources of variability that can generate spurious associations. Batch effects in the genotyping experiment and in the genotype calling also may cause both Type I errors and Type II errors. In our opinion, translation of GWAS findings to clinical practice and personalized medicine and nutrition remains a difficult challenge as we still need to understand the potential sources of spurious results and derive methods to decrease such associations.

Disclaimer

The views presented in this chapter do not necessarily reflect those of the US Food and Drug Administration.

17

Standard Operating Procedures in Clinical Gene Expression Biomarker Panel Development

Khurram Shahzad, Anshu Sinha, Farhana Latif, and Mario C Deng

Abstract

The development of genomic biomarker panels in the context of personalized medicine aims to address biological variation (disease etiology, gender, etc.) while at the same time controlling technical variation (noise). Whether biomarker trials are undertaken by single clinical/laboratory units or in multicenter collaborations, technical noise can be addressed (though not completely eliminated) by the implementation of Standard Operating Procedures (SOPs). Once agreed upon by the study members, SOPs have to become obligatory. We illustrate the usefulness of SOPs by describing the development of a genomic biomarker panel for the absence of acute cellular cardiac allograft rejection resulting from the Cardiac Allograft Rejection Gene Expression Observational (CARGO) study. This biomarker panel is the first Food and Drug Administration-approved genomic biomarker test in the history of transplantation medicine.

17.1 Introduction

In the context of genomics biomarker development, we face in principle two sources of noise – one being technical noise, the other biological variation. The concept of personalized medicine leaves us with modeling large biological patient-to-patient variation (disease etiology, gender, age etc.). Biological noise can only be addressed by strategic sampling from a large variety of individuals. Technical noise, on the other hand, needs to be addressed and minimized by laboratory technological considerations, so that its impact does not diffuse biological or clinical information. The general concept which

Batch Effects and Noise in Microarray Experiments: Sources and Solutions edited by A. Scherer
© 2009 John Wiley & Sons, Ltd

is behind all these considerations is the development of Standard Operating Procedures (SOPs) which define all tasks from design, patient enrollment, and conduct to analysis and interpretation. SOPs are developed before the trial begins.

In the following we describe strategies for minimizing technical and biological/clinical variation by implementation of SOPs in a clinical biomarker study, the Cardiac Allograft Rejection Gene Expression Observational (CARGO) study (Deng *et al.* 2006a). The goal of the CARGO study was the identification of stable heart transplant with a low probability of acute cellular cardiac allograft rejection without performing the protocol endomyocardial biopsy (EMB) which had been the standard of care in rejection surveillance after cardiac transplantation. EMB is an invasive, expensive method which is subject to sampling error and inter-observer variability (Crespo-Leiro *et al.* 2009). Noninvasive methods such as radionuclide imaging had been difficult to validate and implement. The CARGO study group hypothesized that peripheral blood mononuclear cells (PBMCs) may reflect early host response and that measuring PBMC gene expression profiles may yield helpful information with respect to the immune and rejection status of a cardiac transplant recipient. The CARGO clinical trial resulted in one of the few genomics biomarkers that have been approved by the US Food and Drug Administration for commercial use.

17.2 Theoretical Framework

The strategic planning and development of SOPs in clinical gene expression biomarker panel development can best be conceptualized within the framework of systems biology. The completion of the Human Genome Project revolutionized biomedical sciences, providing an answer to the question of the genetic sequence of the human species, yet raising questions about the relationship of the genetic sequence to the proteome and phenome. In order to address these questions, the theoretical framework of systems biology has increasingly been utilized.

17.3 Systems-Biological Concepts in Medicine

Systems-biological approaches include level distinction (also termed multi-scaling), component interaction, emergent properties, downward causation, and dynamic 'feedback loop' behavior. The term 'emergent properties' refers to properties that are present at higher levels of a system, such as heart transplant rejection or multi-organ dysfunction (MOD), but cannot be monocausally explained by the properties of the individual components at a lower system level, such as individual gene transcripts. 'Downward causation' refers to effects of properties at higher levels of the system, such as rejection of the heart or MOD, on the properties at a lower system level, such as individual gene transcripts. Dynamic *feedback loop behavior* is the probabilistic effect of iterative interactions of systems components (e.g. a set of gene transcripts) on the properties of other systems components (e.g. MOD). In essence, the systems-biological approach postulates that quantifiable differential expression patterns of individual genes, proteins, and modules at the molecular and cellular level are simultaneously related, in a mathematically describable probabilistic way, to the clinical entity of interest at the highest systems level, the phenotype level (Figure 17.1). Within the concept of multiscaling of different hierarchical levels, the relationship between properties at the lower system level and emergent properties at

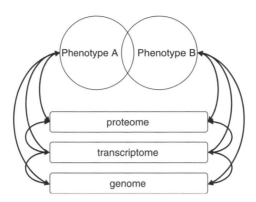

Figure 17.1 Systems-biological research approach in gene expression profiling biomarker development. The phenotype level is related to the proteome level, transcriptome level and genome level (for details, see text).

the higher system level are key to understanding the phenotype within the concept of systems biology.

17.4   General Conceptual Challenges

The major challenge in clinical research and biomarker discovery studies is to control for circumstances unrelated to the relationship between transcriptome and phenome level (Rothman *et al.* 2008). The principle of an SOP is to control for epidemiological, technical, biological/physiological, and statistical variation, which is unrelated to the outcome. This principle is applied to every step of experimental design, conduct, data analysis and interpretation.

Another challenge to the validity of the findings relates to the number of samples to be included in a study (Simon 2005). Michiels *et al.* (2005) published a reanalysis of seven publically available microarray studies which all had the goal of cancer outcome prediction. The authors showed that the list of genes identified as predictors of prognosis was highly unstable, and molecular signatures strongly depended on the selection of the number of samples in the training sets. This highlights the primary challenge in using microarray data, where one is patient-limited and gene-rich: whether genes and signatures are truly significant or whether they are products of random variation, that is, overfitted to noise (Halloran *et al.* 2006). Michiels *et al*, also emphasize the need for determination of the analysis strategy before the actual mathematical process begins, as should be done in an SOP.

17.5   Strategies for Gene Expression Biomarker Development

Based on a clinically important phenotype of interest, gene expression classifier development is conducted in well-defined phases (Table 17.1): (1) consensus operational phenotype definition; (2) candidate gene discovery using a combination of genome-wide and knowledge-base approaches; (3) internal differential gene list validation using PCR assays; (4) diagnostic classifier development using rigorous statistical methods; (5) external clinical validation in a prospective and blinded study; (6) clinical implementation; and (7) post-implementation studies.

Table 17.1   Strategies in gene expression based biomarker test development.

Phase	Tasks	Challenges
Phase 1	CLINICAL PHENOTYPE CONSENSUS DEFINITION	• Imperfect clinical/phenotype standards • Dichotomous vs. continuous phenotype choices
Phase 2	GENE DISCOVERY	• Multicenter study • General gene discovery strategy • Focused vs. whole-genome microarray
Phase 3	INTERNAL DIFFERENTIAL GENE LIST VALIDATION	• RT-PCR validations
Phase 4	DIAGNOSTIC CLASSIFIER DEVELOPMENT	• Discriminatory vs. classifier genes • Mathematical modeling • Biological plausibility of the diagnostic test gene list
Phase 5	EXTERNAL CLINICAL VALIDATION	• Clinical test replicability • Independence of primary clinical validation cohort • Prevalence estimation of clinical phenotype of interest
Phase 6	CLINICAL IMPLEMENTATION	• Regulatory approval • Payer reimbursement • Clinical acceptance
Phase 7	POST-CLINICAL IMPLEMENTATION STUDIES	

17.5.1   Phase 1: Clinical Phenotype Consensus Definition

In medicine any clinical problem may elicit biomarker test development. The consensus definition of the presence and absence of the clinical phenotype of interest is critically important. The expectations of a biomarker test include: identification of patients at high risk for future development of a phenotype of interest; identification of patients in the early stages of the development of phenotype of interest; and differential diagnosis between different phenotypes.

The foundation of a diagnostic test development requires a clinical standard against which to measure the new test performance. The existence of imperfect clinical standards creates a formidable challenge. Often the clinical standard is a biopsy-based histopathological assessment. Biopsy sampling, however, is subject to sampling error, as technical procedures may vary from center to center and also from clinician to clinician. For microarray experiments it is crucial that the procedure is clearly defined. As organs consist of a multitude of different cell types with their own gene expression pattern, great care needs to be taken that the biopsy always represents the same type of tissue and is taken at the same location of an organ for better comparison. In the next step, the histopathological assessment by a trained pathologist is subject to inter-observer variability (Furness *et al.* 2003). For example, in the CARGO study, the derivation of the discriminant equation was critically dependent upon pathological classifications by multiple independent readers. The study design assumed a 'gold' standard clinical endpoint of biopsy-based detection of rejection. However, the CARGO study demonstrated that the clinical standard was limited by considerable inter-observer variability (Marboe *et al.* 2005). To overcome inter-observer and inter-center variability in the CARGO study, biopsies performed by standard techniques were graded by local pathologists and a subset of biopsy samples, including all local grade 1B, 2, 3A and 3B and a representative set of grade 0 and 1A samples, were also graded by three independent ('central') pathologists blinded to clinical information. After an evaluation of the concordance of these biopsy grades by the four pathologists, criteria for selecting acute cellular rejection and quiescent samples were defined prior to developing and validating the classifier. Centralized pathology reading was used to identify these samples, where at least two of four pathologists were required to classify a sample as grade 3A or worse for rejection, and three of four pathologists were required to classify a sample as grade 0 for quiescence. These criteria were set prospectively based upon centralized reading of over 800 CARGO samples and were used in the development and validation of diagnostic polymerase chain reaction (PCR) arrays (Deng *et al.* 2006a). In the CARGO II study, the US–European validation study of Allomap performance outside the US, a similar variability of endomyocardial biopsy reading was confirmed. For a descriptive analysis of core pathology panel interpretation versus local pathologist interpretation we randomly selected a subset of 473 biopsies from 279 cardiac transplant recipients at 12 North American and European centers. Three core pathologists independently reviewed each case for acute cellular rejection (ACR) and were blinded to prior interpretations and any clinical data. Core pathologists confirmed 145 of 227 cases (64%) graded 0, and 9 (37%) of 24 biopsies graded 3A or worse by local pathologists, suggesting substantial inter-reader variability, specifically with regard to grades 3A or worse. These data challenge the reliability of biopsy for post-transplant management (Crespo-Leiro *et al.* 2009).

Dichotomous phenotypes – typically at one end of a spectrum having a clinical picture with fully developed signs/symptoms and at the other end of the spectrum a contrasting picture with no clinical signs/symptoms – are more likely to be associated with differential gene expression signatures and therefore are more appropriate choices for a proof-of-principle study. However, the choice of a continuous phenotype spectrum may allow for a more practical outcome validation approach. If the differential gene expression profiles (GEPs) associated with the extreme phenotypes for the initial proof-of-principle study can be complemented subsequently with intermediate phenotypes and intermediate GEP, a continuous diagnostic test can be developed.

17.5.2   Phase 2: Gene Discovery

After the consensus definition of the – dichotomous or continuous – phenotype of interest, the gene discovery phase is critical.

A multicenter study design increases sample size but also variability. For example, eight US centers were involved in the original CARGO study. Their rejection monitoring protocol varied by (1) biopsy schedule, (2) sampling specimen number, (3) specimen processing and rejection assessment, including antibody-mediated rejection definition, and

(4) immunosuppressive strategies. With respect to multicenter study team organization, it was crucial that every participating physician, technician or nurse was informed about the study vision and strategy. Only trained and motivated personnel are willing to follow specified procedures which may deviate from the process usually followed in the center. In general, sample processing, array processing, and array analysis should be organized by consensus SOPs. An SOP should only be implemented after pilot studies showed their validity. Pilot or feasibility studies can be helpful in the early identification of problems in the collection, handling, and processing of specimens before a larger study is undertaken. These pilot studies may also help in determining the new processes and training requirements before the implementation of a new protocol. The CARGO team paid clear attention that the clinical trial protocol addressed microarray-related aspects of sampling, sample processing, and storage. The SOP which was distributed to the participating centers and which was agreed upon by everyone involved, including clinicians and bioinformaticians, specified in detail how biopsies and blood should be drawn, in which tubes, at what temperature they should be stored, and for how long. Only centers which agreed on these specifications were allowed to participate in the study. In general, this process must take place well before the first sample is taken, and also include each center being informed on the technologies that are applied to their samples. Many specimen collection protocols have special requirements for preservation of macromolecules and/or analytics of interest. In addition to specimen type, things to consider when planning to collect specimens include the collection method, the collection tubes or containers needed, the population that will provide the specimen, personnel required to collect the specimen (and training in the collection process), the distance from the collection point to the processing lab and to the storage facility (if this is a different location), stabilizing or preservation techniques for maintaining/preserving macromolecules required for the specific analyses, and specimen labeling and tracking strategies. For example, in the CARGO study, PBMCs were isolated from 8 mL of venous blood using density gradient centrifugation (CPT, Becton-Dickinson). Samples were frozen in lysis buffer (RLT, Qiagen) within two hours of phlebotomy. Total RNA was isolated from each sample (RNeasy, Qiagen). The effects of processing time on gene assays were tested in PBMCs isolated from six venous blood samples from each of nine donors. Samples were treated identically, except that the interval between blood draw and first centrifugation step was varied from 1 to 8 hours. Any gene arrays showing significant systematic variations across this time period were eliminated from the development process. Importantly, each center was provided with the same materials for tissue handling (Deng *et al.* 2006b). Tissue samples were sent to a core microarray facility, which followed another SOP for each sample. Processing of all samples in a single microarray laboratory, possibly by one technician, limits the technical noise induced by handling of the samples by several individuals or by using different workstations in different laboratories. At the first level of clinical biomarker development one would like to avoid too much unwanted systematic technical noise.

A variety of strategic decisions in the gene discovery process need to be made in the study planning process. For example, in the CARGO study, two approaches were combined, a focused leukocyte microarray and a knowledge-base or literature review, similar to that used to derive a validated PCR-based test for breast cancer recurrence (Paik *et al.* 2004). Our goal was to find a set of genes that could be reproducibly measured by real-time PCR (RT-PCR) in a PBMC RNA preparation. The microarray approach used was limited to genes expressed in leukocytes, potentially ignoring important genome-wide interactions. The knowledge-base approach, focusing on known genes, ignored new biology which might be apparent in a non-hypothesis-driven approach. Although these complementary approaches led to the identification of genes that can distinguish the different rejection states, whole-genome arrays may yield additional, different, or better gene candidates (for a review, see Simon 2005).

17.5.3 Phase 3: Internal Differential Gene List Confirmation

Since the array-based nonsupervised approach to discovering differentially expressed genes is limited by sensitivity and variability, a second molecular biological method such as RT-PCR with higher sensitivity and lower variability needs to be applied to internally validate the discovery phase gene list. For RT-PCR confirmation studies, genes which do not discriminate between absence and presence of the phenotype of interest are considered as normalization genes. Additional assays should be included as controls to detect genomic DNA contamination by the difference between a transcribed and nontranscribed region of a gene. (such as the Gus-B gene) and a spiked-in control template (such as the Arabidopsis gene) to determine if the PCR reaction was successful (Deng *et al.* 2006a).

The differences observed between the microarray and PCR methods in significance of specific genes may be due to: (1) lower sensitivity of microarrays, leading to the elimination of genes which show discrimination in PCR; (2) enhanced reproducibility of RT-PCR, allowing measurement of small differences in gene expression (Palka-Santini *et al.* 2009), likely undetectable by microarrays (usually eliminating genes that show fold changes less than 1.5 to 2). While a complete list validation by RT-PCR is the most complete approach, it is also the most resource-intensive approach. Based on the level of significance (false discovery rate less than 0.05) and the available resources, the gene list can be reduced to a small list for the validation studies.

17.5.4 Phase 4: Diagnostic Classifier Development

The goal for a diagnostic classifier development is the systematic reduction of the validated differentially expressed gene list into a minimum number algorithm which then is subjected to external validation in Phase 5.

One must differentiate between the identification of discriminatory genes and the development of a robust classifier. For example, in the CARGO study, several cytotoxic T-cell genes were not selected for the classifier by the automated and unbiased statistical method of linear discriminant analysis, while PDCD1, a known marker of T-cell activation, whose expression is correlated with the other T-cell genes, was chosen in the final model. The fact that some of the 'well-known' markers of this pathway were not selected by the method points to the strong relevance of PDCD1 as the optimal representative of this pathway and suggests that these other genes were either not incrementally informative or lacked sufficient reproducibility to be included in the model.

Classifier development poses unique mathematical modeling challenges. For example in the CARGO study, for discriminant equation development purposes, RT-PCR data on 252 genes for the training set of 36 rejection and 109 quiescent CARGO samples were generated to derive a panel of candidate genes for classifier development and to validate microarray results. Gene expression results were analyzed with Student's t-test, median ratios, hierarchical clustering by TreeView and an expert assessment of biological relevance. Metagenes, defined as transcripts behaving in a concordant manner (Brunet et al. 2004), were constructed by averaging gene expression levels that were correlated across training samples with correlation coefficients of at least 0.7. Genes significantly distinguishing rejection from quiescence in the PCR training set by t-test ($p \leq 0.01$), by median ratio differences less than 0.75 or greater than 1.25, or by correlation with significant genes were used for metagene construction and classifier development. The methods for analyzing gene expression data included principal component analysis, linear discriminant analysis (LDA, StatSoft Inc.), logistic regression (SAS Institute Inc.), prediction analysis of microarrays (PAM), voting, classification and regression trees (TreeNet, Salford Systems), random forests, nearest shrunken centroids, and k-nearest neighbor. We sought to develop a classifier that quantitatively distinguished current moderate/severe acute cellular rejection (International Society for Heart and Lung Transplantation (ISHLT) grade 3A or worse) from quiescence (ISHLT grade 0) using gene and metagene expression levels as the variables. The final classifier was developed using LDA as implemented in the Discriminant Function Analysis module of Statistica (StatSoft Inc.). LDA constructs a linear classifier by automatically selecting genes and/or metagenes that, in combination, optimally separate rejection and quiescent samples in the training set. The robustness of selected genes, and the appropriate number of genes in the classifier were both evaluated by cross-validation.

An important challenge lies in understanding how genomic results correlate to known or 'expected' biological pathways for the disease state. Biological plausibility is a valid criterion to increase one's confidence in a genomic result, but the converse – the lack of such information – is *not* evidence that a genomic result is false since the biological literature is not complete in this regard. In fact, genomic technologies enable researchers to discover new genes or pathways associated with disease that may not be expected from the existing literature. If only 'expected' genes are accepted in genomics studies, new discovery will be stunted (Deng et al. 2006b).

17.5.5   Phase 5: External Clinical Validation

The external clinical validation of the constructed diagnostic classifier in an independent patient sample is critically important in GEP-biomarker development.

During the development of a gene expression classifier we always encounter uncertainty about clinical reproducibility of findings. Additional studies using different patient cohorts are recommended to further support and extend initial validation results. Rigorous mathematical modeling and receiver operator characterstic (ROC) curve analysis are conducted to construct a diagnostic test score range (e.g. 0–40 for the CARGO classifier) with the best possible cutoff in this range which rules out phenotypes of interest with high negative predictive value (e.g. moderate to severe rejection in CARGO). Area under the curve (AUC) analysis can be used as a criterion to estimate how well the classifier performs in internal training and test sets of samples. In addition, to establish test reproducibility and to demonstrate acceptable precision of the test results, further studies can be designed to study the effect of different sources of variation within the sample processing which may cause noise in test results. Such sources of variation are, for example, operator-to-operator variation, run-to-run variation, lot-to-lot variation of reagents, plate-to-plate variation, and section-to-section variation of the plates used to run the test. Hence, it is imperative to validate these initial results in independent, external patient samples.

In the CARGO study, an independent cohort of CARGO patients was selected to validate the effectiveness of the LDA classifier defined in the diagnostic development phase using a prospective and blinded study protocol. The primary objective of the validation study was to test the pre-specified hypothesis that the diagnostic score distinguishes quiescence, defined as ISHLT grade 0 from moderate/severe biopsy-proven acute rejection, defined as ISHLT grade 3A or worse, both grades determined by local and centralized cardio-pathological examination. This was assessed using a two-tailed Student's t-test for comparing score distributions for rejection and quiescent samples. Secondary and exploratory objectives included documentation of diagnostic performance across thresholds and description of correlations to clinical variables.

Results for the validation study are reported for unique samples from patients not used for training (primary validation study), as well as for a larger set of samples not used for training (secondary validation study). These latter samples may provide improved power but may be biased to the extent that a longitudinal set of samples from an individual patient is not independent with respect to gene expression. To estimate prevalence of the clinical phenotype of interest, for example in the CARGO study, a representative set of samples, across all local biopsy grades and at least 1 year post-transplant were evaluated to assess the discriminant equation performance on a stable patient population. From these samples, positive predictive value (PPV: fraction of samples with scores at or above the threshold expected to have concurrent biopsy grade 3A or worse) and negative predictive value (NPV: fraction of samples with scores below the threshold expected to be free from biopsy grade 3A or worse) were estimated at multiple test thresholds. Given the risk associated with undetected acute cellular rejection, and the clinical use of EMB, we sought a threshold that maximized the NPV at the expense of the PPV.

17.5.6   Phase 6: Clinical Implementation

The clinical implementation of a gene expression biomarker based diagnostic test requires regulatory approval, payer reimbursement, and clinical acceptance.

While diagnostic tests in the pre-genome era were based on known biology, the situation in the post-genome era is different. Diagnostic tests contain genes, the biological role of which is not fully known, while their predictive performance is well established. For regulatory approval the FDA recently created the new category of in-vitro-diagnostic multivariate index assay (IVDMIA) (http://www.fda.gov/cdrh/oivd/meetings/020807/Williams.html). The ability to generate genetic profiles using microarrays and sequence-based approaches promises to expand greatly the utility of genetic tests in clinical medicine. This is because human diseases resulting from a single genetic alteration are rare. Most common diseases including cancer, cardiovascular disease, and diabetes result from a variety of genetic changes acting in concert. Moreover, the exact combination of genetic factors resulting in a specific disease often varies among individuals. To address this complexity, many companies and academic groups are developing complex molecular diagnostics (including IVDMIA tests based on microarrays) with the goal of establishing correlations between a specific pattern of genetic modification and/or gene expression and disease outcomes such as progression, response to therapy, or adverse reactions. In some cases, these correlations and their predictive value are strong enough that the tests can have clinical utility even in the absence of a full understanding of the effects of and interactions among the component genes (http://www.ostp.gov/galleries/PCAST/pcast_report_v2.pdf).

For unanimous payer reimbursement decisions, different levels of approaches have been applied including the California Technology Assessment Forum's Technology Assessment criteria. These include the following criteria: (1) The technology must have the appropriate regulatory approval. (2) The scientific evidence must permit conclusions concerning the effectiveness of the technology regarding health outcomes. (3) The technology must improve the net health outcomes. (4) The technology must be as beneficial as any established alternatives. (5) The improvement must be attainable outside the investigational setting. The clinical acceptance is based on complete evidence, safety/efficacy profile, cost/benefit profile and availability.

17.5.7   Phase 7: Post-Clinical Implementation Studies

Validation must occur at two levels. The first level is confirmation that the correlation initially observed, whether through a genomics-based population study or a study based on biospecimen, is indeed real rather than a statistical artifact. This level of validation is achieved by repeating the preliminary analysis on an independent population or biospecimen sample set as outlined for the Phase 6 study tasks. The second level of validation is to confirm that use of a diagnostic test based on the correlation actually results in improved clinical outcomes for patients. This definitive validation requires a prospective clinical trial. Once the clinical implementation phase commences, ongoing observational studies based on the clinical experience at different centers will help to define the safety/efficacy profiles and normal application aspects.

Full clinical implementation of novel tests in the IVDMIA category is challenging. The Invasive Monitoring Attenuation Through Gene Expression (IMAGE) study is a prospective, multicenter, nonblinded randomized clinical trial designed to test the hypothesis that a primarily noninvasive rejection surveillance strategy utilizing GEP testing is noninferior to an invasive EMB-based strategy with respect to cardiac allograft dysfunction, rejection with hemodynamic compromise, and all-cause mortality. A total of 500 heart transplant recipients who are in their second through fifth post-transplant years are currently being enrolled in the IMAGE study between 2005 and 2009. The IMAGE study is the first randomized controlled comparison of two rejection surveillance strategies measuring outcomes in heart transplant recipients who are beyond their first year post-transplant. Moving away from routine histologic evaluation for allograft rejection represents an important paradigm shift in cardiac transplantation, and the results of this study have important implications for the future management of heart transplant patients (Pham *et al.* 2007).

17.6 Conclusions

The implementation of strict SOPs which acknowledge and address aspects of microarray analysis for each step of the process from histopathology through sampling and microarray processing to data handling is essential to biomarker development. Only the rigorous control of each step can ensure an accomodation although not an elimination of technical noise. SOPs are essential components of the experimental design phase in clinical single- or multicenter biomarker development studies. Following good SOPs ensures the minimization of unwanted impact of factors which are nonessential for the outcome being investigated. SOPs determine each step from design to patient enrollment, through conduct, to interpretation. It is essential to report each step in a transparent fashion, including any deviations from the SOP. SOPs need to be agreed upon by the entire team, and only centers and teams which agree and which are motivated to follow the SOP will participate in the study, ensuring the best possible data quality for successful biomarker discovery. Examples of guidelines on how to write SOPs and enhance thoroughness and transparency of reporting are given in United States Environmental Protection Agency (2007), Bossuyt *et al.* (2003), and Clive (2004).

18

Data, Analysis, and Standardization

Gabriella Rustici, Andreas Scherer, and John Quackenbush

Abstract

'Reporting noise' is generated when data and their metadata are described, stored, and exchanged. Such noise can be minimized by developing and adopting data reporting standards, which are fundamental to the effective interpretation, analysis and integration of large data sets derived from high-throughput studies. Equally crucial is the development of experimental standards such as quality metrics and a consensus on data analysis pipelines, to ensure that results can be trusted, especially in clinical settings. This chapter provides a review of the initiatives currently developing and disseminating computational and experimental standards in biomedical research.

18.1 Introduction

Noise is unavoidable in any measurement and must be understood if those measurements are to be effectively used in any scientific analysis. There are two major sources of noise in biological experiments: biological noise and technical noise. Biological noise is intrinsic to living systems and consists, in its most basic form, of cell-to-cell variability; it can be estimated and accounted for by carefully designing experiments that include replicates of the treatments or conditions being tested (see Chapters 3 and 4 in the present volume). Technical noise, on the other hand, is introduced at each step of a laboratory procedure and consists, *inter alia*, of differences in sample preparation, in processing, and of the uncertainties inherent in any measurement. It too can be estimated, modelled, and corrected (see various chapters in this volume).

In addition, noise is generated when one collects, describes, formats, submits, and exchanges data and information related to such data; we refer to this as 'reporting noise'.

Batch Effects and Noise in Microarray Experiments: Sources and Solutions edited by A. Scherer
© 2009 John Wiley & Sons, Ltd

Reporting noise reflects inaccuracies that occur either through errors such as mislabeled data and transposed rows in a file or due to incomplete descriptions of experimental variables necessary for the analysis and interpretation of the data. Reporting noise can be minimized by developing and adopting data reporting standards, which are sets of guidelines agreed across communities for describing, publishing and disseminating experimental data.

Data standards are of fundamental importance in high-throughput research; as scientists try to understand and reconstruct the interactions between different components of biological systems using genomic, transcriptomic, proteomic, and metabolomic approaches (the 'omics'), the integration of data generated from such technologies requires high levels of standardization. Together with computational standards, experimental standards such as quality controls, best practices, and consensus on key elements in data analysis pipelines are also needed to facilitate data integration. Finally, rigorous methods for the assessment of data quality are required to ensure that results can be trusted, especially in the context of clinical applications.

This chapter provides an overview of ongoing initiatives dedicated to developing and disseminating both computational and experimental standards for high-throughput technologies in biomedical research. All the websites mentioned herein are listed in Table 18.1.

18.2 Reporting Standards

In the last two decades, the fields of biomedical and biological research have undergone remarkable changes as they have transitioned from gene-centered to genome-scale sciences – a transition that is moving biology from a purely laboratory science to what is increasingly an information science. Novel omic technologies have revolutionized the traditional hypothesis-driven approach to scientific investigation and are now nearly indispensable tools in biomedical research. Consequently, biology has become a very heterogeneous discipline facing the outstanding challenge of integrating knowledge derived from different omic disciplines with the ultimate goal of describing and modeling the structure and behavior of a biological system as a whole (Ideker *et al.* 2001).

The rapid expansion of these high-throughput methodologies and the vast amount of raw data generated by them has been a source of both tremendous opportunities and significant informatics challenges. New computational tools that enable data analysis and knowledge capture across heterogeneous sources of information are constantly under development. More importantly, omic data generally come with a complex set of ancillary information that is ultimately essential for effective data analysis. Commonly referred to as 'metadata', this information ideally contains a formalized description of a data set's content (sample characteristics, experimental design, protocols, ...). Because it is necessary for analysis, these metadata need to be carefully recorded and stored with the primary omic data (Quackenbush 2004).

Because of the costs and challenges in assembling genomic data from large populations, sharing data effectively is essential, but analyzing them requires an understanding of and the use of the relevant metadata. Consequently, data have to be: (i) reported and described in an exhaustive, unambiguous and standardized way in order to capture all

Table 18.1   Useful websites.

Resource	URL
ABRF	http://www.abrf.org/sPRG
ArrayExpress	http://www.ebi.ac.uk/microarray-as/ae/at
BII	http://isatab.sourceforge.net/tools.html
Bioconductor	http://www.bioconductor.org/
Biomodels.net	http://biomodels.org/
Biomodels	http://www.ebi.ac.uk/biomodels-main/
BioPAX	http://www.biopax.org/
BIRN	http://www.nbirn.net/
caBIG	https://cabig.nci.nih.gov/
CDISC	http://www.cdisc.org/standards/index.html
CellML	http://www.cellml.org/models/
CIBEX	http://cibex.nig.ac.jp/index.jsp
CLSI	http://www.clsi.org/
EMERALD	http://www.microarray-quality.org/
Equator	http://www.equator-network.org/
FUGE	http://fuge.sourceforge.net
GeneRIF	http://www.ncbi.nlm.nih.gov/projects/GeneRIF/
Gene Wiki portal	http://en.wikipedia.org/wiki/Portal:Gene_Wiki
GEO	http://www.ncbi.nlm.nih.gov/geo/
GO	http://www.geneontology. org
GSC	http://gensc.org/gsc/
HL7	http://www.hl7.org/
Human Proteinpedia	http://www.humanproteinpedia.org/
HUPO	http://www.hupo.org
ISAcreator	http://isatab.sourceforge.net/isacreator.html
ISA-TAB format	http://isatab.sourceforge.net
KEGG	http://www.genome.jp/kegg/
MAQC	http://www.fda.gov/nctr/science/centers/toxicoinformatics/maqc/
MGED	http://www.mged.org
MIBBI	http://www.mibbi.org/
NCBI Taxonomy database	http://www.ncbi.nlm.nih.gov/Taxonomy/
NIST	http://www.nist.gov/
OBI (ex-FuGO)	http://obi-ontology.org/
OBO Foundry	http://www.obofoundry.org/
OLS	http://www.ebi.ac.uk/ontology-lookup/
Phenote	http://www.phenote.org/
PRIDE	http://www.ebi.ac.uk/pride/
Proteome Harvest	http://www.ebi.ac.uk/pride/proteomeharvest/
PSI	http://www.psidev.info
Reactome	http://www.reactome.org
RSBI	http://www.mged.org/Workgroups/rsbi/
SBO	http://www.ebi.ac.uk/sbo/
SEND	http://www.cdisc.org/models/send/v2.3
TPA	http://www.ncbi.nlm.nih.gov/Genbank/TPA.html
WikiPathways	http://www.wikipathways.org/
WikiProteins	http://www.wikiprofessional.org

relevant information about an experiment; and (ii) stored in public databases to facilitate data sharing among the scientific community. After all, as elegantly quoted in a recent *Nature Cell Biology* editorial, 'the value of any data is only as good as its annotation and accessibility' (Nature Cell Biology 2008).

What all of this suggests is that data reporting standards must be developed to facilitate comparison and integration of omic data sets; and new approaches for describing, storing, and exchanging omic data formats are required for publishing scientific results and making them quickly and reliably accessible.

Data standardization efforts focus on answering three major questions:

1. What information do we need to capture?
2. What syntax (or file format) should we use to exchange data?
3. What semantics (or ontology) should we use to best describe the metadata?

To answer these questions, various standardization initiatives have been developing data reporting standards, including the following:

1. Checklists, which outline the minimum information (MI) content that should be specified when reporting both data and metadata. These vary from general guidelines to itemized lists of information that should be provided whenever describing an experiment so that, in the future, an independent scientist will be able to carry out an independent analysis of the data and reproduce the same results. One challenge here is that our understanding of what information is 'minimal' is constantly evolving.
2. File formats, which facilitate the exchange of data and related information. The MI requirements are conceptualized and implemented through the creation of an object model (OM), which reflects the structure of the data and is then translated into a communication format for data exchange, often implemented using an eXtensible Markup Language (XML)-based format. Here the challenge is often reflected in balancing the needs of making the annotation user-friendly and computer-readable.
3. Ontology, which is a formal specification of terms in a particular subject area and the relations among them. Its purpose is to provide a basic, stable and unambiguous description of such terms and relations in order to avoid improper and inconsistent use of the terminology pertaining to a given domain. Thus far, Gene Ontology (GO) has been the most successful ontology initiative (Ashburner *et al.* 2000). GO is a controlled vocabulary used to describe the biology of a gene product in any organism. Each gene is assigned three independent terms describing: (i) the molecular function of the gene product; (ii) the biological process in which the gene product participates; and (iii) the cellular component where the gene product can be found. The identification of enriched GO terms among clusters of co-regulated genes has become a routine step in the analysis of gene expression data. To be successful, one must create an ontology that is well defined yet reasonable in size but that does not overly limit our ability to describe an experiment.

Several organizations and committees are tackling data standardization, focusing on various aspects. Here we will present an overview of initiatives that have emerged from the academic community and that address standards-related issues relating to particular technologies and applications including genomics, transcriptomics, proteomics, metabolomics and systems biology.

18.3   Computational Standards: From Microarray to Omic Sciences

18.3.1   The Microarray Gene Expression Data Society

No technology embodies the rise of omic science more than DNA microarrays. Since their appearance in the early 1990s, microarrays have become a common tool for expression, genetic and epigenetic studies, and have started to enter the more demanding domain of clinical applications and diagnostics.

The Microarray and Gene Expression Data Society (MGED), established in 1999, has been the pioneer in developing guidelines for describing and storing microarray data. MGED's efforts resulted in the creation of:

1. the Minimum Information About a Microarray Experiment (MIAME) checklist and standards for microarray data annotation (Brazma *et al.* 2001). MIAME is the most mature standards initiative and is considered the benchmark for the development of new standards.
2. the MicroArray Gene Expression Object Model (MAGE-OM) and its XML-based implementation (MAGE-ML; Spellman *et al.* 2002), designed to describe and communicate information about microarray-based experiments. The complexity of MAGE-ML made it difficult for biologists to use it and a simpler spreadsheet format, MAGE-TAB, is now recommended for representing microarray data and metadata (Rayner *et al.* 2006).
3. the MGED Ontology, as an annotation resource for microarray data (Whetzel *et al.* 2006b).

The MIAME standards have been widely adopted by the microarray community and are supported by end users, vendors and journal editors. Most scientific journals now require that all experimental data be placed in a public repository and that the relevant accession numbers be provided as a condition for publication. In support of MIAME and the associated publication requirements, MGED worked with the three primary public biological data repositories and their gene expression databases (Ball *et al.* 2004): Array-Express (Brazma *et al.* 2003), Gene Expression Omnibus (GEO; Edgar *et al.* 2002) and CIBEX (Ikeo *et al.* 2003).

As gene expression analysis became more widely applied, extensions to MIAME were developed for plant biology (MIAME/Plant; Zimmermann *et al.* 2006), nutrigenomics (MIAME/Nutr), toxicogenomics (MIAME/Tox; Sansone *et al.* 2004) and environmental biology (MIAME/Env). The last three have been promoted and coordinated by the Reporting Structure for Biological Investigations MGED working group (RSBI) which has focused on developing cross-discipline, interoperable standards to facilitate integration of data from multiple omic technologies (Sansone *et al.* 2006). Guidelines are also being developed for transcriptomic applications based on high-throughput sequencing – Minimum Information about a high-throughput Nucleotide SeQuencing Experiment (MIN-SEQE). Similar initiatives have now spread to other omic sciences, including proteomics, metabolomics, genomics and systems biology.

18.3.2   The Proteomics Standards Initiative

Within the Human Proteome Organization (HUPO), the Proteomics Standards Initiative (PSI) has developed reporting requirements for proteomic experiments, including file formats and controlled vocabularies for the exchange of proteomic data (Taylor *et al.* 2006). Given the diversity of proteomic technologies covered by PSI, there are six working groups, each contributing to:

1. the Minimum Information About a Proteomic Experiment reporting guidelines (MIAPE; Taylor *et al.* 2007). MIAPE is organized into modules specific to a particular technology or group of technologies, as well as general modules that are relevant to all experimental designs.
2. the technology-specific data exchange formats including the Molecular Interaction Format (MIF; Hermjakob *et al.* 2004) for the exchange of molecular interaction data and the mzData format for mass spectrometry.
3. the controlled vocabularies providing a consensus annotation system to standardize the meaning, syntax and formalism of terms used across proteomics. Each working group is developing the controlled vocabulary required by the particular technology or data type, adopting the same standardization for overlapping concepts.

The PSI-compliant Proteomics Identifications database (PRIDE) aims to become the proteomic equivalent of gene expression databases and serve as a public repository of proteomic data supporting journal publications (Jones *et al.* 2008). Several other databases exist for storing proteomic data from mass spectrometry and a collaborative agreement on data sharing between them is currently underway (Hermjakob and Apweiler 2006).

18.3.3   The Metabolomics Standards Initiative

The Metabolomics Standards Initiative (MSI) was recently established (Sansone *et al.* 2007) but is working closely with MGED and PSI to rapidly develop protocols for standardization and reporting of metabolomic analyses. MSI has five working groups that are involved in developing:

1. Core Information for Metabolomics Reporting (CIMR), which will specify the minimum reporting guidelines for metabolomic work;
2. Ontology, a controlled vocabulary developed on the basis of CIMR; and
3. Exchange format, a data model and exchange format developed on the basis of CIMR, making use of the MSI ontology.

Currently, there is no public repository for storing data from metabolomic experiments.

18.3.4   The Genomic Standards Consortium

Standards for the description of genomes and the exchange and integration of genomic data are being developed by the Genomic Standards Consortium (GSC), which is bringing together researchers, bioinformaticians, computer scientists, and ontology experts dealing with genomic information. This community has recently introduced the Minimum Information about a Genome Sequence (MIGS; Field *et al.* 2008), for describing collection of genomes and metagenomes acquired by both traditional and next generation sequencing methods and is now working towards the development of the Genomic Contextual Data Markup Language (GCDML; Kottmann *et al.* 2008).

18.3.5 Systems Biology Initiatives

The computational systems biology community has been dealing with the need for standards for quantitative biochemical model generation and curation. Until now, the majority of published quantitative models in biology have been lost as they are unavailable to the wider community or insufficiently characterized to be reusable.

The BioModels.net initiative has established a framework of rules, known as the Minimal Information Requirements In the Annotation of Models (MIRIAM), to favour collection of high-quality models and facilitate their curation (Le Novere *et al.* 2005). Models must be encoded in the Systems Biology Markup Language (SBML; Hucka *et al.* 2003) or CellML (Lloyd *et al.* 2004), and the Systems Biology Ontology (SBO) must be used to annotate models with connections to biological data resources (Le Novere 2006).

Several public repositories provide free and easy access to annotated models; these include BioModels (Le Novere *et al.* 2006) and the CellML repository (Lloyd *et al.* 2008).

Collaborative efforts are also underway for the development of a data exchange format and an ontology for biological pathway data. The Biological Pathways Exchange (BioPAX) data exchange format facilitates information retrieval from multiple pathway databases and integration of four main categories of pathway data: metabolic pathways, molecular interactions, gene regulation networks, and signaling pathways (Luciano 2005). The BioPAX ontology makes use of existing controlled vocabularies, such as GO, OBO (see below), and the NCBI taxonomy database, and is designed to support the data models of a number of existing pathway databases, including KEGG (Kanehisa *et al.* 2004) and Reactome (Vastrik *et al.* 2007).

18.3.6 Data Standards in Biopharmaceutical and Clinical Research

Governmental regulatory organizations, such as the US Food and Drug Administration (FDA), have been promoting the development of data standards for clinical applications with particular emphasis on best practices for collecting and reporting clinical trial data (Frueh 2006). The FDA has also been one of the major contributors in the establishment of the Clinical Data Interchange Standards Consortium (CDISC), which has now emerged as a leading voice for data standardization within the biopharmaceutical and clinical research communities. The CDISC has so far developed a set of standards supporting all aspects of the drug development lifecycle, including study design, collection, exchange, analysis, reporting, and archiving of clinical and nonclinical data (Souza *et al.* 2007). These include the Study Data Tabulation Model (SDTM) which has been approved by the FDA as the recommended standard for submitting clinical trial data for regulatory submissions.

Standardization efforts have now moved beyond the original goal of establishing a simple data interchange standard and are now focusing on defining a controlled terminology to ensure standards interoperability. This is possibly the greatest challenge for the biomedical community, considering the wide variety of terms used to express similar or identical concepts (Kush *et al.* 2008); such inconsistency renders vain any attempt at data comparison, integration, and sharing among different organizations, greatly reducing clinical research effectiveness. A partnership has been established between CDISC and the National Cancer Institute (NCI) Enterprise Vocabulary Services to standardize medical terminology between industry and academia.

CDISC is now working closely with several primary healthcare organizations, including Health Level 7 (HL7), in the attempt to harmonize clinical research and healthcare standards (Kush *et al.* 2008). These harmonization efforts are still in their infancy and the data standards scenario for clinical applications remains highly fragmented. In a recently compiled inventory of contributors, more than 50 organizations and 100 distinct standards were identified (https://www.ctnbestpractices.org/standards-inventory/).

18.3.7   Standards Integration Initiatives

The domain-specific initiatives described thus far have been of great importance in addressing issues related to data capture requirements. However, as one might imagine, the emergence of multiple standards has resulted in a fragmented, redundant, and often conflicting patchwork set of requirements. More significantly, the development of different data exchange formats and ontologies limits the overall benefits of data standardization and seriously jeopardizes data integration. Ideally, standards should be self-sufficient but also compatible with other standards to better support work that increasingly draws on multiple large-scale sources of data to fully drive understanding of the relevant biological problems under investigation (Sansone and Field 2006).

Some synergistic activities, involving the functional genomics and systems biology communities, have been established with the aim of harmonizing and consolidating reporting standards across domains. The challenges that these new initiatives face include removing redundancies, resolving conflicting standards, filling gaps between domains that are covered by existing standards, developing compatible file exchange formats and ontologies, and creating tools for data integration. All this must be accomplished while supporting existing standards, allowing them to evolve, and ensuring both forward and backward compatibility.

Among these highly interconnected initiatives are:

- the Minimal Information for Biological and Biomedical Investigations (MIBBI) project that is promoting gradual integration of MI requirements;
- the Open Biomedical Ontologies (OBO) Foundry that is delivering a set of common terms for the integration of controlled vocabularies;
- the Functional Genomics Experiment (FuGE) model and ISA-TAB format that seek to represent data derived from diverse experimental techniques using a common framework; and

• the BioInvestigation Index infrastructure that aims to store experimental metadata and sample-data relationships from multi-omic studies.

18.3.8 The MIBBI project

MIBBI provides a point of reference in the fast proliferating field of data standards initiatives. Through its internet 'Portal', it offers a freely accessible resource which allows users to browse all of the reporting 'checklists' registered with it (Taylor *et al.* 2008). For each MIBBI-affiliated project, summary information and updates of the checklist developmental status are available, in a simple and straightforward way. As of February 2009, 29 projects have joined this initiative and more are expected to follow (see Table 18.2). Furthermore, MIBBI is involved in promoting checklist integration through its 'Foundry', a community-driven effort examining ways to reduce redundancy between existing checklists and develop a suite of orthogonal standards. MIBBI is strongly related to the Equator Network, which seeks to standardize the reporting of clinical trials in the medical literature.

18.3.9 OBO Foundry

The OBO Foundry is devoted to harmonizing life-science ontologies (Smith *et al.* 2007). As of February 2009, the OBO Foundry comprises 80 ontologies, functioning as an ontology information resource, and it is trying to reduce redundancy among existing ontologies and to align ontology development efforts carried out by separate communities. Software tools using OBO ontologies for annotation purposes are currently under development; they include Phenote, for phenotype annotation, and Proteome Harvest and ISAcreator (see below), for metadata annotation.

Within the OBO Foundry, the Ontology for Biomedical Investigation consortium (OBI; Whetzel *et al.* 2006a) is a large collaborative project, initiated in 2005, which aims to build an integrated ontology for the description of biological and clinical investigations. This will include a set of 'universal' terms, applicable across biological and technological domains, as well as domain-specific terms. OBI will provide terms that can be used to annotate experimental metadata, as well as the data generated and the analysis performed.

18.3.10 FuGE and ISA-TAB

The Functional Genomics Experiment (FuGE) initiative aims to create tools, XML formats and database schemas for capturing data in any biological domain (Jones *et al.* 2007). FuGE offers a single, generic data capture format (FuGE-ML) that allows description of any biomedical science workflow focusing on the common components of experimental annotation and generally applicable information about the design of investigations. In addition, it provides mechanisms for supplementing existing data formats and a framework for building new ones with a common structure for techniques with special requirements.

Table 18.2 MI projects registered with MIBBI.

Checklist acronym	Checklist name	Main website
CIMR	Core Information for Metabolomics Reporting	http://msi-workgroups.sourceforge.net/
MIABE	MI About a Bioactive Entity	TBA
MIACA	MI About a Cellular Assay	http://miaca.sourceforge.org/
MIAME	MI About a Microarray Experiment	www.mged.org/miame
MIAME/Env	MI about an ENvironmental transcriptomic experiment	http://nebc.nox.ac.uk/miame/miame_env.html
MIAME/Nutr	MI about a Nutrigenomics experiment	http://www.mged.org/Workgroups/rsbi/rsbi.html
MIAME/Plant	MI About a Microarray Experiment involving Plants	http://www.ebi.ac.uk/at-miamexpress
MIAME/Tox	MI about an array-based toxicogenomics experiment	http://www.mged.org/Workgroups/rsbi/rsbi.html
MIAPA	MI About a Phylogenetic Analysis	TBA
MIAPAR	MI About a Protein Affinity Reagent	http://www.psidev.info/index.php?q=node/277
MIAPE	MI About a Proteomics Experiment	http://www.psidev.info/miape/
MIARE	MI About an RNAi Experiment	http://www.miare.org
MIASE	MI About a Simulation Experiment	http://www.ebi.ac.uk/compneur-srv/miase/
MIENS	MI about an ENvironmental Sequence	http://gensc.org/gc_wiki/index.php/MIENS
MIFlowCyt	MI for a Flow Cytometry Experiment	http://flowcyt.sourceforge.net/
MIGen	MI about a Genotyping Experiment	
MIGS	MI about a Genome Sequence	http://gensc.sf.net

MIMIx	MI about a Molecular Interaction Experiment	http://psidev.sf.net
MIMPP	MI for Mouse Phenotyping Procedures	http://www.interphenome.org
MINI	MI about a Neuroscience Investigation	www.carmen.org.uk
MINIMESS	Minimal Metagenome Sequence Analysis Standard	TBA
MINSEQE	MI about a high-throughput SeQuencing Experiment	http://www.mged.org/minseqe/
MIPFE	MI for Protein Functional Evaluation	http://mipfe.cogentech.it/
MIQAS	MI for QTLs and Association Studies	http://miqas.sourceforge.net/
MIqPCR	MI about a quantitative Polymerase Chain Reaction experiment	http://www.rdml.org
MIRIAM	MI Required In the Annotation of biochemical Models	http://biomodels.net/index.php?s=MIRIAM
MISFISHIE	MI Specification For In Situ Hybridization and Immunohistochemistry Experiments	http://mged.sourceforge.net/misfishie/
STRENDA (1A & 1B)	1A: Required Data for the Methods Section for Publishing of Enzyme Kinetic Data	http://www.strenda.org/
	1B: Additional Information Required for Reporting Enzyme Kinetic Data	
TBC	Tox Biology Checklist	TBA

A complete set of interoperable FuGE-based modules for multi-omic experiments is still under development and several groups, including members of the standards initiatives described above, have developed a simpler framework, the Investigation/Study/Assay tab-delimited format (ISA-TAB), to describe, submit, and exchange complex metadata from experiments employing multiple omic technologies, along with conventional methodologies (Sansone *et al.* 2008). Where experiments include clinical or nonclinical studies, ISA-TAB can complement existing biomedical formats such as the SDTM, which encompasses both the Standard for Exchange of Non-clinical Data (SEND) and the CDISC, by formally capturing information about the interrelationship of the various parts they describe. Once FuGE-based modules or other interoperable XML formats become available, ISA-TAB can continue to serve those with little or no bioinformatics support, as well as finding utility as a user-friendly presentation layer for XML-based formats.

The ISA-TAB format also provides the basis for the BioInvestigation Index (BII), a new prototype infrastructure for storing experimental metadata and sample-data relationships from multi-omic studies. Public repositories of transcriptomic and proteomic data, such as ArrayExpress and PRIDE, are independent entities, which employ different metadata representations, submission/exchange formats and terminologies. BII addresses the need for a uniform mean of collecting and presenting complex data sets, by leveraging on synergistic standards. Its infrastructure provides the user with: (i) a query interface for browsing and retrieving experimental metadata; and (ii) a standalone annotation/submission tool, named ISAcreator. ISAcreator supports the use of OBO ontologies for metadata annotation, accessible through the European Bioinformatics Institute (EBI) Ontology Lookup Service, and is configured in compliance with existing minimum requirements, according to the relevant MIBBI checklists.

18.4   Experimental Standards: Developing Quality Metrics and a Consensus on Data Analysis Methods

The importance of standardization becomes even more evident when we consider the role of high-throughput technologies in clinical studies and the potential revolution that personalized medicine could bring along, when a patient genotype or expression profile could be used to tailor medical care to an individual's needs. If the decision-making process, for example in the context of clinical trials or prognostic prediction, has to be based on the outcome of an omic experiment, we need to make sure that the results can be trusted. This requires the introduction of quality controls, best practices, quality metrics, and consensus on data analysis pipelines in order to reduce the variability intrinsic to these technologies.

Let us consider the significant changes gene expression microarrays have undergone since the mid 1990s. Despite continued technology development and improvement, reproducibility at different sites and comparability of results obtained on different platforms are still issues; there is a lack of consensus on best practices for experimental design and sample processing, as well as for data analysis and interpretation, reflecting the wide range of applications for the technology. Therefore, more complex experimental standards need to be established and adopted by the scientific community in order to overcome these problems.

Several initiatives are currently underway to improve confidence in the data generated by array experiments, including:

- initiatives aiming to develop quality metrics and quality control procedures, as well as a consensus on data analysis, such as MAQC and EMERALD; and
- initiatives aiming to develop standard reference material for array experiments, such as the External RNA Controls Consortium (ERCC).

Large regulatory bodies such as the FDA, the Environmental Protection Agency (EPA), and the Clinical and Laboratory Standards Institute (CLSI) were the first to raise concern about the issues of data quality assessment and reproducibility (Dix *et al.* 2006; Frueh 2006). The growing need to improve array data quality and its assessment resulted in the formation of the FDA-led MicroArray Quality Control consortium (MAQC), a community-wide effort bringing together researchers from the government, industry and academia seeking to experimentally identify the key factors contributing to variability and reproducibility of microarray data.

Phase I of the MAQC project, launched in 2005 and recently completed, assessed the performance of seven microarray platforms in profiling the expression of two commercially available RNA samples, at multiple test sites (Shi *et al.* 2006). The data set generated was used to compare performances of different data analysis methods in identifying differentially expressed genes. MAQC's main conclusions confirm that microarray results were generally repeatable within test site, reproducible between test sites, if care is taken in using standard practices, and comparable across platforms; these findings corroborate results that had been previously established by other groups (Bammler *et al.* 2005; Irizarry *et al.* 2005; Larkin *et al.* 2005; Kuo *et al.* 2006). The expression patterns generated were a reflection of biology regardless of the differences in technology. Most interestingly, the study showed that one statistical solution cannot be applied to all cases and that the statistical analysis required for clinical diagnostics is likely to be different than those commonly used in basic research.

The MAQC project has now entered phase II, focusing on capabilities and limitations of different data analysis methods for clinical and preclinical microarray applications. The main objective is now to reach a consensus on best practices in developing and validating microarray-based predictive models for clinical and toxicological applications (Shi *et al.* 2008). The best analytical practices identified by MAQC-II will be implemented in a Data Analysis Protocol, a collection of data analysis guidelines applicable to any data set. Results of the study have been submitted for publication and are currently under review.

Related projects have been established within the European Union Sixth Framework Programme for disseminating quality metrics, microarray standards and best laboratory practices. This is the scope of EMERALD, the European Project on Standards and Standardisation of Microarray Technology and Data Analysis. Among the activities of the seven EMERALD working groups is the development of:

1. normalization and transformation ontologies to capture data pre-processing information;
2. experimental standards for controls, experimental design, and execution; and
3. tools for quality assessment.

To clearly demonstrate the quality of data generated from a microarray experiment, objective quality metrics need to be developed. This will improve confidence in conclusions drawn from such studies and facilitate data comparison. For this reason, scientists from the EMERALD consortium have recently released a Bioconductor package, named ArrayQualityMetrics, that provides an HTML report with diagnostic plots for one- or dual-color microarray data (Kauffmann *et al.* 2009). This report can be used as the first step in the microarray analysis pipeline or to compare the efficiency of different pre-processing methods.

Complementary initiatives for developing appropriate microarray reference material that can be used to verify technical performance of gene expression assays are also in progress. The External RNA Control Consortium has been testing and selecting polyadenylated transcripts that can be added to each RNA sample before processing to monitor and evaluate technical performance of both microarrays and real-time reverse-transcription polymerase chain reaction analysis (Baker *et al.* 2005). Approximately 100 external RNA controls have been selected and are now in a final testing phase. Once this phase is completed, the National Institute of Standards and Technology will advance the controls through its Standards Program to make the set available to researchers around the world as a standardized DNA reference material (Salit 2006).

The necessity of developing experimental standards is now being addressed by other omic sciences. The proteomic community is developing simple and complex mixtures of known proteins to confidently assess and validate existing and new experimental procedures, as well as technological, analytical and computational developments (Keller *et al.* 2002; Purvine *et al.* 2004). The Proteomics Standards Research Group of the Association of Biomolecular Resource Facilities promotes and supports the development and use of standards in proteomics. It has created a prototype standard reference which is now being tested with various methodologies for protein identification in mass spectrometry (Hogan *et al.* 2006).

In the future, consortiums similar to MAQC and the External RNA Control Consortium need to be established for the purpose of developing a universal set of standardized DNA references with known genotypes and gene-copy number alterations for use in high-throughput genotyping and sequencing (Ji and Davis 2006). These would provide a valuable tool for the validation of new genomic technologies, a need which the advent of next generation DNA sequencing technologies has made even more imperative.

18.5   Conclusions and Future Perspective

All the standardization initiatives described in this chapter will prove successful only if accepted and embraced by the research community. Active community participation is fundamental to ensuring effective development of standards and their correct use and application. Still researchers too often fail to understand the importance of such initiatives and do not take advantage of them, and funding agencies have been slow to support their development. The broader scientific community needs to fully appreciate how the adoption of such standards can increase the value of their data and the knowledge derived from their findings, and funders need to more completely understand how developing and implementing standards can help them gain additional value from their investment in omic projects (Brooksbank and Quackenbush 2006).

As technologies and their applications continue to develop at a rapid pace, community involvement is the only feasible option for keeping standardization efforts current and relevant. Several projects are already focusing on involving the community in updating the annotation of current data sets. The NCBI Third Party Annotation (TPA) database allows users to annotate and revise records deposited in the nucleic acid sequence databanks GenBank, EMBL and DDBJ. The NCBI also provides the Gene Reference into Function (GeneRIF) tool, allowing scientists to add a concise function annotation to genes described in Entrez Gene.

Wikimedia technologies can be a very effective way to achieve community annotation. Open public platforms, such as the Gene Wiki portal (Huss *et al.* 2008), WikiProteins (Mons *et al.* 2008), Human Proteinpedia (Mathivanan *et al.* 2008) and WikiPathways (Pico *et al.* 2008), are allowing the research community to directly contribute to curating and annotating existing genes, proteins and pathways records, which would otherwise remain static and consequently result in a great loss of scientific information. These projects have only recently been started and it remains to be seen how widely they will be embraced by the biology community (Nature Cell Biology 2008).

Facing a similar challenge are the large scientific 'cyberinfrastructures' such as the Cancer Biomedical Informatics Grid (caBIG) and the Biomedical Informatics Research Network (BIRN). These initiatives aim to establish virtual communities for facilitating data sharing and knowledge exchange, providing scientists with tools for collecting, managing and analyzing data. Despite being the initiative that most embodies the ideal of data integration, caBIG lacks 'grassroots' support from researchers and experimentalists and consequently fails to deliver its great potential (Stein 2008).

Ultimately we believe that standards are important and that their implementation and support are essential for the future of large-scale omic science. As the number of success stories that involve using and reusing data continues to grow, we believe that the broader community and government and private funding agencies will see the value of standards efforts. We hope this review provides the starting point for a broader community discussion regarding standards and have established a forum at http://network.nature.com/groups/data_standardization to allow for community comment and input.

References

Adams, MD, Kelley, JM, Gocayne, JD, *et al.* (1991) Complementary DNA sequencing: Expressed sequence tags and human genome project. *Science*, **252**, 1651–1656.

Alizadeh, AA, Eisen, MB, Davis, RE, *et al.* (2000) Distinct types of diffuse large B-cell lymphoma identified by gene expression profiling. *Nature*, **403**(6769), 503–511.

Allison, DB, Cui, X, Page, GP, *et al.* (2006) Microarray data analysis: from disarray to consolidation and consensus. *Nature Review Genetics*, **7**, 55–65.

Almon, RR, Yang, E, Lai, W, *et al.* (2008) Circadian variations in rat liver gene expression: relationships to drug actions. *Journal of Pharmacology and Experimental Therapeutics*, **326**, 700–716.

Alter, O, Brown, O, and Botstein, O (2000) Singular value decomposition for genome-wide expression data processing and modeling. *Proceedings of the National Academy of Sciences of the USA*, **97**, 10101–10106.

Altman, DG, and Bland, JM (1999) Treatment allocation in controlled trials: why randomize. *British Medical Journal*, **318**, 1209.

Altman, NS (2005) Replication, variation and normalization in microarray experiments. *Applied Bioinformatics*, **4**, 33–44.

Altman, NS, and Hua, J (2006) Extending the loop design for two-channel microarray experiments. *Genetical Research*, **88**, 153–163.

Altshuler, D, and Daly, M (2007) Guilt beyond a reasonable doubt. *Nature Genetics*, **39**, 813–815.

Ambroise, C, and McLachlan, GJ (2002) Selection bias in gene extraction on the basis of microarray gene-expression data. *Proceedings of the National Academy of Sciences of the USA*, **99**, 6562–6566.

Anguiano, A, Nevins, JR, and Potti, A (2008) Toward the individualization of lung cancer therapy. *Cancer*, **113**(Suppl.), 1760–1767.

Ashburner, M, Ball, C, Blake, J, *et al.* (2000) Gene ontology: tool for the unification of biology. The Gene Ontology Consortium. *Nature Genetics*, **25**, 25–29.

Bakay, M, Chen, YW, Borup, R, *et al.* (2002) Sources of variability and effect of experimental approach on expression profiling data interpretation. *BMC Bioinformatics*, **3**, 4.

Baker, SC, Bauer, SR, Beyer, RP, *et al.* (2005) The External RNA Controls Consortium: a progress report. *Nature Methods*, **2**, 731–734.

Baldi, P, and Long, AD (2001) A Bayesian framework for the analysis of microarray expression data: regularized *t*-test and statistical inferences of gene changes. *Bioinformatics*, **17**, 509–519.

Ball, C, Brazma, A, Causton, H, *et al.* (2004) An open letter on microarray data from the MGED Society. *Microbiology*, **150**, 3522–3524.

Bammler, T, Beyer, RP, Bhattacharya, S, *et al.* (2005) Standardizing global gene expression analysis between laboratories and across platforms. *Nature Methods*, **2**, 351–356.

Barrett, JC, and Cardon, LR (2006) Evaluating coverage of genome-wide association studies. *Nature Genetics*, **38**, 659–662.

Bengtsson, H, Jonsson, G, and Vallon-Christersson, J (2004) Calibration and assessment of channel-specific biases in microarray data with extended dynamical range. *BMC Bioinformatics*, **5**, 177.

Batch Effects and Noise in Microarray Experiments: Sources and Solutions edited by A. Scherer
© 2009 John Wiley & Sons, Ltd

Benito, M, Parker, J, Du, Q, *et al.* (2004) Adjustment of systematic microarray data biases. *Bioinformatics*, **20**, 105–114.

Benjamini, Y, and Hochberg, Y (1995) Controlling the false alarm discovery rate: a practical and powerful approach to multiple testing. *Journal of the Royal Statistical Society, Series B*, **57**, 289–300.

Boedigheimer, MJ, Wolfinger, RD, Bass, MB, *et al.* (2008) Sources of variation in baseline gene expression levels from toxicogenomics study control animals across multiple laboratories. *BMC Genomics*, **9**, 285–300.

Boelens, MC, Meerman, GJ, Gibcus, JH, *et al.* (2007) Microarray amplification bias: loss of 30% differentially expressed genes due to long probe – poly(A)-tail distances. *BMC Genomics*, **8**, 277.

Bolstad, BM, Irizarry, RA, Astrand, M, *et al.* (2003) A comparison of normalization methods for high density oligonucleotide array data based on variance and bias. *Bioinformatics*, **19**, 185–193.

Bolton, S, and Bon, C (2004) *Pharmaceutical Statistics. Practical and Clinical Applications*, 4th edn. New York: Marcel Dekker.

Bossuyt, PM, Reitsma, JB, Bruns, DE (2003) The STARD statement for reporting studies of diagnostic accuracy: explanation and elaboration. *Annals of Internal Medicine*, **138**, W1–12.

Bottinger, EP, Ju, W, and Zavadil, J (2002) Applications for microarrays in renal biology and medicine. *Experimental Nephrology*, **10**, 93–101.

Box, GEP, Hunter, JS, and Hunter, WG (2005) *Statistics for Experimenters. Design, Innovation, and Discovery*, 2nd edn. Hoboken, NJ: John Wiley & Sons, Inc.

Branham, WS, Melvin, CD, Han, T, *et al.* (2007) Elimination of laboratory ozone leads to a dramatic improvement in the reproducibility of microarray gene expression measurements. *BMC Biotechnology*, **7**, 8.

Brazma, A, Hingamp, P, Quackenbush, J, *et al.* (2001) Minimum information about a microarray experiment (MIAME) – toward standards for microarray data. *Nature Genetics*, **29**, 365–371.

Brazma, A, Parkinson, H, Sarkans, U, *et al.* (2003) ArrayExpress – a public repository for microarray gene expression data at the EBI. *Nucleic Acids Research*, **31**, 68–71.

Brem, RB, Yvert, G, Clinton, R, *et al.* (2002) Genetic dissection of transcriptional regulation in budding yeast. *Science*, **296**, 752–755.

Brettschneider, J, Bolstad, BM, Collin, F, *et al.* (2008) Quality assessment for short oligonucleotide microarray data. *Technometrics*, **50**, 241–283.

Brooksbank, C, and Quackenbush, J (2006) Data standards: a call to action. *Omics*, **10**, 94–99.

Brown, PO, and Botstein, D (1999) Exploring the new world of the genome with DNA microarrays. *Nature Genetics*, **21**(Suppl.), 33–37.

Brunet, JP, Tamayo, P, Golub, TR, *et al.* (2004) Metagenes and molecular pattern discovery using matrix factorization. *Proceedings of the National Academy of Sciences of the USA*, **101**, 4164–4169.

Buch, S, Schafmayer, C, Völzke, H, *et al.* (2007) A genome-wide association scan identifies the hepatic cholesterol transporter ABCG8 as a susceptibility factor for human gallstone disease. *Nature Genetics*, **39**, 995–999.

Bylesjö, M, Eriksson, D, Sjodin, A, *et al.* (2007) Orthogonal projections to latent structures as a strategy for microarray data normalization. *BMC Bioinformatics*, **8**(1), 207.

Cardon, LR, and Palmer, LJ (2003) Population stratification and spurious allelic association. *Lancet*, **361**, 598–604.

Chanderbali, AS, Albert, VA, Leebens-Mack, J, *et al.* (2009) Transcriptional signatures of ancient floral developmental genetics in avocado (*Persea americana; Lauraceae*). *Proceedings of the National Academy of Sciences of the USA*, **106**, 8929–8934.

Chen, JJ, Hsueh, HM, Delongchamp, RR, *et al.* (2007) Reproducibility of microarray data: a further analysis of MicroArray Quality Control (MAQC) data. *BMC Bioinformatics*, **8**, 412.

Chen, Y, Dougherty, ER, and Bittner, ML (1997) Ratio-based decisions and the quantitive analysis of cdna microarray images. *Journal of Biomedical Optics*, **2**, 364–374.

Chomczynski, P, and Sacchi, N (1987) Single-step method of RNA isolation by acid guanidinium thiocyanate-phenol-chloroform extraction. *Analytical Biochemistry*, **162**, 156–159.

Chu, T, Glymour, C, Scheines, R, *et al.* (2003) A statistical problem for inference to regulatory structure from associations of gene expression measurements with microarrays. *Bioinformatics*, **19**, 1147–1152.

Churchill, GA (2002) Fundamentals of experimental design for cDNA microarray. *Nature Genetics*, **32**(Suppl.), 490–495.

Churchill, GA (2004) Using ANOVA to analyze microarray data. *Biotechniques*, **37**, 173–175, 177.

Cleveland, WS (1979) Robust locally weighted regression and smoothing scatterplots. *Journal of the American Statistical Association*, **74**, 829–836.

Clive, CM (2004) *Handbook of SOPs for Good Clinical Practice*, 2nd edn. Boca Raton, FL: Interpharm/CRC Press.

Cobb, K (2006) Microarrays: The search for meaning in a vast sea of data. *Biomedical Computation Review*, **2**(6), 16–23.

Cochran, WG, and Cox, GM (1962) *Experimental Design*. New York: John Wiley & Sons, Inc..

Coombes, KR, Highsmith, WE, Krogmann, TA, *et al.* (2002) Identifying and quantifying sources of variation in microarray data using high-density cDNA membrane arrays. *Journal of Computational Biology*, **9**, 655–669.

Crespo-Leiro, MG, Schulz, U, Vanhaecke, J, *et al.* (2009) Inter-observer variability in the interpretation of cardiac biopsies remains a challenge: Results of the Cardiac Allograft Rejection Gene Expression Observational (CARGO) II Study (abstr.). *Journal of Heart and Lung Transplantation*, **28**(Suppl. 1), S230.

Cui, X, and Churchill, GA (2003a) Statistical tests from differential expression in cDNA microarray experiments. *Genome Biology*, **4**, 210–219.

Cui, X, and Churchill, GA (2003b) How many mice and how many arrays? Replication in mouse cDNA microarray experiments. In SM Lin and KF Johnson (eds), *Methods of Microarray Data Analysis III*. Norwell, MA: Kluwer Academic Publishers.

Culhane AC, Perrière, G, and Higgins, DG (2003) Cross-platform comparison and visualization of gene expression data using co-inertia analysis. *BMC Bioinformatics*, **4**, 59.

Daskalakis, A, Kostopoulos, S, Spyridonos, P, *et al.* (2008) Design of a multi-classifier system for discriminating benign from malignant thyroid nodules using routinely H&E-stained cytological images. *Computers in Biology and Medicine*, **38**, 196–203.

Day, SJ, and Altman, DG (2000) Blinding in clinical trials and other studies. *British Medical Journal*, **321**, 504.

Dempster, AP, Laird, N, and Rubin, DB (1977) Maximum likelihood from incomplete data via the EM algorithm. *Journal of the Royal Statistical Society, Series B*, **39**, 1–38.

Deng, MC, Eisen, HJ, Mehra, M. R., *et al.* (2006a) Noninvasive discrimination of rejection in cardiac allograft recipients using gene expression profiling. *American Journal of Transplantation*, **6**, 150–160.

Deng, MC, Eisen, HJ, and Mehra, MR (2006b) Methodological challenges of genomic research – the CARGO study. *American Journal of Transplantation*, **6**, 1086–1087.

Dewan, A, Liu, M, Hartman, S, *et al.* (2006) HTRA1 promoter polymorphism in wet age-related macular degeneration. *Science*, **314**, 989–992.

DeRisi, JL, Iyer, VR, and Brown, PO (1997) Exploring the metabolic and genetic control of gene expression on a genomic scale. *Science*, **278**, 680–686.

Di, X, Matsuzaki, H, Webster, TA, *et al.* (2005) Dynamic model based algorithms for screening and genotyping over 100 K SNPs on oligonucleotide microarrays. *Bioinformatics*, **21**, 1958–1963.

Dix, DJ, Gallagher, K, Benson, WH, *et al.* (2006) A framework for the use of genomics data at the EPA. *Nature Biotechnology*, **24**, 1108–1111.

Dobbin, KK, and Simon, R (2002) Comparison of microarray designs for class comparison and class discovery. *Bioinformatics*, **18**, 1438–1445.

Dobbin, KK, Shih, JH, and Simon, R (2003) Statistical design of reverse dye microarrays. *Bioinformatics*, **19**, 803–810.

Dobbin, KK, Beer, DG, Meverson, M, *et al.* (2005a) Interlaboratory comparability study of cancer gene expression analysis using oligonucleotide microarrays. *Clinical Cancer Research*, **11**, 565–572.

Dobbin, KK, Kawasaki, ES, Petersen, DW, *et al.* (2005b) Characterizing dye bias in microarray experiments. *Bioinformatics*, **21**, 2430–2437.

Dudley, AM, Aach, J, Steffen, MA, *et al.* (2002) Measuring absolute expression with microarrays with a calibrated reference sample and an extended signal intensity range. *Proceedings of the National Academy of Sciences of the USA*, **99**, 7554–7559.

Dudoit, S, Yang, YH, Callow, MJ, *et al.* (2002) Statistical methods for identifying differentially expressed genes in replicated cDNA microarray experiments. *Statistica Sinica*, **12**(1), 111–140.

Edgar, R, Domrachev, M, and Lash, AE (2002) Gene Expression Omnibus: NCBI gene expression and hybridization array data repository. *Nucleic Acids Research*, **30**, 207–210.

Edwards, DE (2003) Non-linear normalization and background correction in one-channel cDNA microarray studies. *Bioinformatics*, **19**, 825–833.

Efron, B, Tibshirani, R, Storey, JD, *et al.* (2001) Empirical Bayes analysis of a microarray experiment. *Journal of the American Statistical Association*, **96**, 1151.

Ein-Dor, L, Zuk, O, and Domany, E (2006) Thousands of samples are needed to generate a robust gene list for predicting outcome in cancer. *Proceedings of the National Academy of Sciences of the USA*, **103**, 5923–5928.

Eisen, MB (1999) ScanAlyze, User Manual. http://rana.lbl.gov/manuals/ScanAlyzeDoc.pdf.

Eklund, AC, and Szallasi, Z (2008) Correction of technical bias in clinical microarray data improves concordance with known biological information. *Genome Biology*, **9**, p. R26.

Fan, C, Oh, DS, Wessels, LB, *et al.* (2006) Concordance among gene-expression-based predictors for breast cancer. *New England Journal of Medicine*, **355**, 560–569.

Fare, TL, Coffey, EM, Dai, H, *et al.* (2003) Effects of atmospheric ozone on microarray data quality. *Analytical Chemistry*, **75**, 4672–4675.

Feng, S, Wolfinger, RD, Chu, TM, *et al.* (2006) Empirical Bayesian Analysis of Variance Component Models for Microarray Data. *Journal of Agricultural, Biological, and Environmental Statistics*, **11**, 197–209.

Field, D, Garrity, G, Gray, T, *et al.* (2008) The minimum information about a genome sequence (MIGS) specification. *Nature Biotechnology*, **26**, 541–547.

Fisher, RA (1938) Presidential Address: The First Session of the Indian Statistical Congress. *Sankhyā: The Indian Journal of Statistics*, **4**, 14–17.

Fodor, SP, Read, JL, Pirrung, MC, *et al.* (1991) Light-directed, spatially addressable parallel chemical synthesis. *Science*, **251**, 767–773.

Frantz, S (2005) An array of problems. *Nature Reviews Drug Discovert*, **4**, 362–363.

Frayling, TM, Timpson, NJ, Weedon, MN, *et al.* (2007) A common variant in the FTO gene is associated with body mass index and predisposes to childhood and adult obesity. *Science*, **316**, 889–894.

Frueh, FW (2006) Impact of microarray data quality on genomic data submissions to the FDA. *Nature Biotechnology*, **24**, 1105–1107.

Furness, PN, Taub, N, Assmann, KJ, *et al.* (2003) International variation in histologic grading is large, and persistent feedback does not improve reproducibility. *American Journal of Surgical Pathology*, **27**, 805–810.

Gilks, WR, Richardson, S, and Spiegelhalter, DJ (1996) Introduction. In WR Gilks, S Richardson, and DJ Spiegelhalter (eds), *Markov Chain Monte Carlo in Practice*. London: Chapman & Hall.

Gold, B, Kirchhoff, T, Stefanov, S, *et al.* (2008) A genome-wide association study provides evidence for a breast cancer risk at 6q22.33. *Proceedings of the National Academy of Sciences of the USA*, **105**, 4340–4345.

Gold, LS, Manley, NB, Slone, TH, *et al.* (1999) Supplement to the Carcinogenic Potency Database (CPDB): results of animal bioassays published in the general literature in 1993 to 1994 and by the National Toxicology Program in 1995 to 1996. *Environmental Health Perspectives*, **107**(Suppl. 4), 527–600.

Gottardo, R, Raftery, AE, Yee Yeung, K *et al.* (2005) Bayesian robust inference for differential gene expression in microarrays with multiple samples. *Biometrics*, **62**, 10–18.

Grimm, LG, and Yarnold, PR (2001) *Reading and Understanding Multivariate Analysis*. Washington, DC: American Psychological Association.

Gunderson, KL, Kruglyak, S, Graige, MS, *et al.* (2004) Decoding randomly ordered DNA arrays. *Genome Research*, **14**, 870–877.

Guo, L, Lobenhofer, EK, Wang, C, *et al.* (2006) Rat toxicogenomic study reveals analytical consistency across microarray platforms. *Nature Biotechnology*, **24**, 1162–1169.

Hacia, JG, Fan, JB, Ryder, O, *et al.* (1999) Determination of ancestral alleles for human single-nucleotide polymorphisms using high-density oligonucleotide arrays. *Nature Genetics*, **22**, 164–167.

Hakonarson, H, Grant, SF, Bradfield, JP, *et al.* (2007) A genome-wide association study identifies KIAA0350 as a type 1 diabetes gene. *Nature*, **448**, 591–594.

Hall, P, Marron, JS, and Neeman, A, (2005) Geometric representation of high dimension low sample size data. *Journal of the Royal Statistical Society, Series B*, **67**, 427–444.

Halloran, PF, Reeve, J, and Kaplan, B (2006) Lies, damn lies, and statistics: the perils of the *p* value. *American Journal of Transplantation*, **6**, 9–11.

Han, ES, Wu, Y, McCarter, R, *et al.* (2004) Reproducibility, sources of variability, pooling, and sample size: important considerations for the design of high-density oligonucleotide array experiments. *Journals of Gerontology Series A: Biological Sciences and Medical Sciences*, **59**, 306–315.

Hawkins, DM (2004) The problem of overfitting. *Journal of Chemical Information and Computer Sciences*, **44**, 1–12.

Heidecker, B, and Hare, JM (2007) The use of transcriptomic biomarkers for personalized medicine. *Heart Failure Reviews*, **12**, 1–11.

Hermjakob, H, and Apweiler, R (2006) The Proteomics Identifications Database (PRIDE) and the ProteomExchange Consortium: making proteomics data accessible. *Expert Reviews of Proteomics*, **3**, 1–3.

Hermjakob, H, Montecchi-Palazzi, L, Bader, G, *et al.* (2004) The HUPO PSI's molecular interaction format – a community standard for the representation of protein interaction data. *Nature Biotechnology*, **22**, 177–183.

Hirschhorn, JN, and Daly, MJ (2005) Genome-wide association studies for common disease and complex traits. *Nature Review Genetics*, **6**, 95–108.

't Hoen, PAC, de Kort, F, van Ommen, GJB, *et al.* (2003) Fluorescent labelling of cRNA for microarray Applications. *Nucleic Acids Research*, **31**, e20.

Hogan, JM, Higdon, R, and Kolker, E (2006) Experimental standards for high-throughput proteomics. *Omics*, **10**, 152–157.

Hong, H, Su, Z, Ge, W, *et al.* (2008) Assessing batch effects of genotype calling algorithm BRLMM for the Affymetrix GeneChip Human Mapping 500K Array Set using 270 HapMap samples. *BMC Bioinformatics*, **9**(Suppl.), S17.

Howson, JM, Barratt, BJ, Todd, JA, *et al.* (2005) Comparison of population- and family-based methods for genetic association analysis in the presence of interacting loci. *Genetic Epidemiology*, **29**, 51–67.

Huang, J, Qi, R, Quackenbush, J, *et al.* (2001) Effects of ischemia on gene expression. *Journal of Surgical Research*, **99**, 222–227

Hu, R, Qiu, X, Glazko, G, *et al.* (2009) Detecting intergene correlation changes in microarray analysis: a new approach to gene selection. *BMC Bioinformatics*, **10**, 1.

Hucka, M, Finney, A, Sauro, HM, *et al.* (2003) The systems biology markup language (SBML): a medium for representation and exchange of biochemical network models. *Bioinformatics*, **19**, 524–531.

Hughes, TR, Mao, M, Jones, AR, *et al.* (2001) Expression profiling using microarrays fabricated by an ink-jet oligonucleotide synthesizer. *Nature Biotechnology*, **19**, 342–347.

Huss, JW, 3rd, Orozco, C, Goodale, J, *et al.* (2008) A gene wiki for community annotation of gene function. *PLoS Biology*, **6**, p. e175.

Ideker, T, Galitski, T, and Hood, L (2001) A new approach to decoding life: systems biology. *Annual Review of Genomics and Human Genetics*, **2**, 343–372.

Ikeo, K, Ishi-i, J, Tamura, T, *et al.* (2003) CIBEX: center for information biology gene expression database. *Comptes Rendues Biologies*, **326**, 1079–1082.

ILSI Health and Environmental Sciences Institute (2003) ILSI HESI Technical Committee on the Application of Genomics to Mechanism-Based Risk Assessment. http://dels.nas.edu/emergingissues/docs/Pettit.pdf.

International HapMap Consortium, Frazer, KA, Ballinger, DG, Cox, DR, *et al.* (2007) A second generation human haplotype map of over 3.1 million SNPs. *Nature*, **449**, 851–862.

Ioannidis, JPA (2005) Microarray and molecular research: noisy discovery. *Lancet*, **365**, 454–455.

Irizarry, RA, Hobbs, B, Collin, F, *et al.* (2003) Exploration, normalization, and summaries of high density oligonucleotide array probe level data. *Biostatistics*, **4**, 249–264.

Irizarry, RA, Warren, D, Spencer, F, *et al.* (2005) Multiple-laboratory comparison of microarray platforms. *Nature Methods*, **2**, 345–350.

ISO (2007) *International Vocabulary of Metrology – Basic and General Concepts and Associated Terms (VIM)*, ISO/IEC Guide 99:2007. Geneva: International Organization for Standardization http://www.bipm.org/utils/common/documents/jcgm/JCGM_200_2008.pdf.

Ji, H, and Davis, RW (2006) Data quality in genomics and microarrays. *Nature Biotechnology*, **24**, 1112–1113.

Jin, W, Riley, RM, Wolfinger, RD, *et al.* (2001) The contributions of sex, genotype and age to transcriptional variance in drosophila melanogaster. *Nature Genetics*, **29**, 389–395.

Johnson, RA, and Wichern DW (2002) *Applied Multivariate Statistical Analysis*, 5th edn. Upper Saddle River, NJ: Prentice Hall.

Johnson, WE, Li, C, and Rabinovic, A (2007) Adjusting batch effects in microarray expression data using empirical Bayes methods. *Biostatistics*, **8**, 118–127.

Jones, AR, Miller, M, Aebersold, R, *et al.* (2007) The Functional Genomics Experiment model (FuGE): an extensible framework for standards in functional genomics. *Nature Biotechnology*, **25**, 1127–1133.

Jones, P, Cote, RG, Cho, SY, *et al.* (2008) PRIDE: new developments and new datasets. *Nucleic Acids Research*, **36**, D878–883 (Database issue).

Juenger, TE, Wayne, T, Boles, S, *et al.* (2006) Natural genetic variation in whole-genome expression in *Arabidopsis thaliana*: the impact of physiological QTL introgression. *Molecular Ecology*, **15**, 1351–1365.

Kacew, S (2001) Confounding factors in toxicity testing. *Toxicology*, **160**, 87–96.

Kanehisa, M, Goto, S, Kawashima, S, *et al.* (2004) The KEGG resource for deciphering the genome. *Nucleic Acids Research*, **32**, D277–280 (Database issue).

Karssen, AM, Li, JZ, Her, S, *et al.* (2006) Application of microarray technology in primate behavioral neuroscience research. *Methods*, **38**, 227–234.

Katz, S, Irizarry, RA, Lin, X, *et al.* (2006) A summarization approach for Affymetrix GeneChip data using a reference training set from a large, biologically diverse database. *BMC Bioinformatics*, **7**, 464.

Kauffmann, A, Gentleman, R, and Huber, W (2009) arrayQualityMetrics – a bioconductor package for quality assessment of microarray data. *Bioinformatics*, **25**, 415–416.

Keller, A, Purvine, S, Nesvizhskii, AI, *et al.* (2002) Experimental protein mixture for validating tandem mass spectral analysis. *Omics*, **6**, 207–212.

Kendziorski, CM, Zhang, Y, Lan, H, and Attie, AD (2003) The efficiency of mRNA pooling in microarray experiments.. *Biostatistics*, **4**, 465–477.

Kendziorski, C, Irizarry, RA, Chen, KS, *et al.* (2005) On the utility of pooling biological samples in microarray experiments. *Proceedings of the National Academy of Sciences of the USA*, **102**, 4252–4257.

Kerr, MK (2003) Design considerations for efficient and effective microarray studies. *Biometrics*, **59**, 822–828.

Kerr, MK, and Churchill, GA (2001a) Experimental design for gene expression microarrays. *Biostatistics*, **2**, 183–201.

Kerr, MK, and Churchill, GA (2001b) Statistical design and the analysis of gene expression microarray data. *Genetical Research*, **77**, 123–128.

Kerr, MK, Afshari, C, Bennett, L, *et al.* (2000a) Statistical analysis of a gene expression microarray experiment with replication. *Statistica Sinica*, **12**, 203–217.

Kerr, MK, Martin, M, and Churchill, GA (2000b) Analysis of variance for gene expression microarray data. *Journal of Computational Biology*, **7**, 819–837.

Kita, Y, Shiozawa, M, Jin, W, *et al.* (2002) Implications of circadian gene expression in kidney, liver and the effects of fasting on pharmacogenomic studies. *Pharmacogenetics*, **12**, 55–65.

Klebanov, L, and Yakovlev, A (2007) Diverse correlation structures in gene expression data and their utility in improving statistical inference. *Annals of Applied Statistics*, **1**(2), 538–559.

Klebanov, L, and Yakovlev, A (2008) A nitty-gritty aspect of correlation and network inference from gene expression data. *Biology Direct*, **3**, 1–30.

Klebanov, L, Qiu, X, Welle, S, and Yakovlev, A (2007) Statistical methods and microarray data. *Nature Biotechnology*, **25**, 25–26.

Klebanov, L, Qiu, X, and Yakovlev, A (2008) Testing differential expression in nonoverlapping gene pairs: a new perspective for the empirical Bayes method. *Journal of Bioinformatics and Computational Biology*, **6**, 301–316.

Kooperberg, C, Fazzio, TG, Delrow, JJ, *et al.* (2002) Improved background correction for spotted DNA microarrays. *Journal of Computational Biology*, **9**, 55–66.

Kottmann, R, Gray, T, Murphy, S, *et al.* (2008) A standard MIGS/MIMS compliant XML schema: toward the development of the Genomic Contextual Data Markup Language (GCDML). *Omics*, **12**, 115–121.

Kuehl, RO (2000) *Design of Experiments: Statistical Principles of Research Design and Analysis*, 2nd revised edn. Pacific Grove, CA: Duxbury/Thomson Learning.

Kuo, WP, Jenssen T-K, Butte AJ, *et al.* (2002) Analysis of matched mRNA measurements from two different microarray technologies. *Bioinformatics*, **18**, 405–412.

Kuo, WP, Liu, F, Trimarchi, J, *et al.* (2006) A sequence-oriented comparison of gene expression measurements across different hybridization-based technologies. *Nature Biotechnology*, **24**, 832–840.

Kush, RD, Helton, E, Rockhold, FW, *et al.* (2008) Electronic health records, medical research, and the Tower of Babel. *New England Journal of Medicine*, **358**, 1738–1740.

Lamb, J, Crawford, ED, Peck, D, *et al.* (2006) The Connectivity Map: using gene-expression signatures to connect small molecules, genes, and disease. *Science*, **313**, 1929–1935.

Lander, ES (1999) Array of hope'. *Nature Genetics*, **21**, 3–4.

Lander ES., Linton LM., Birren B, *et al.* (2001) Initial sequencing and analysis of the human genome. *Nature*, **409**, 860–921.

Larkin, JE, Frank, BC, Gavras, H, *et al.* (2005) Independence and reproducibility across microarray platforms. *Nature Methods*, **2**, 337–344.

Le Novere, N (2006) Model storage, exchange and integration. *BMC Neuroscience*, **7**(Suppl. 1), S11.

Le Novere, N, Finney, A, Hucka, M, *et al.* (2005) Minimum information requested in the annotation of biochemical models (MIRIAM). *Nature Biotechnology*, **23**, 1509–1515.

Le Novere, N, Bornstein, B, Broicher, A, *et al.* (2006) BioModels Database: a free, centralized database of curated, published, quantitative kinetic models of biochemical and cellular systems. *Nucleic Acids Research*, **34**, D689–691 (Database issue).

Lebl, M, Burger, C, Ellman, B, *et al.* (2001) Fully Automated Parallel Oligonucleotide Synthesizer. *Collection of Czechoslovak Chemical Communications*, **66**(8), 1299–1314

Lee, CH, and Macgregor, PF (2004) Using microarrays to predict resistance to chemotherapy in cancer patients. *Pharmacogenomics*, **5**, 611–625.

Lee, KM, Kim, JH, and Kang, D (2005) Design issues in toxicogenomics using DNA microarray experiment. *Toxicology and Applied Pharmacology*, **207**(2, Suppl. 1), 200–208.

Lee, ML, Lu, W, Whitmore, GA, *et al.* (2002) Models for microarray gene expression data. *Journal of Biopharmaceutical Statistics*, **12**, 1–19.

Lee, MLT, Kuo, FC, Whitmore, GA, *et al.* (2000) Importance of replication in microarray gene expression studies: statistical methods and evidence from repetitive cDNA hybridizations. *Proceedings of the National Academy of Sciences of the USA*, **97**, 9834–9839.

Leek, JT., and Storey, JD. (2007) Capturing heterogeneity in gene expression studies by surrogate variable analysis. *PLoS Genetics*, **3**, 1724–1735.

Leo, WR, 1994, *Techniques for Nuclear and Particle Physics Experiments: A How-to Approach*. New York: Springer-Verlag.

Li, C, and Wong, W (2003) DNA-Chip Analyzer (d-Chip). In G Parmigiani, ES Garrett, RA Irizarry, and SL Zeger (eds), *The Analysis of Gene Expression Data: Methods and Software*. New York: Springer.

Li, M, Boehnke, M, and Abecasis, GR (2006) Efficient study designs for test of genetic association using sibship data and unrelated cases and controls. *American Journal of Human Genetics*, **78**, 778–792.

Li, X, Gu, WMS, and Balink, D (2002) DNA Microarrays: their use and misuse. *Microcirculation*, **9**, 13–22.

Liang, P, and Pardee, AB (1992) Differential display of eukaryotic messenger RNA by means of the polymerase chain reaction. *Science*, **257**, 967–971.

Liggett, W (2008) Technical variation in the modeling of the joint expression of several genes. In P Stafford (ed.), *Methods in Microarray Normalization*. Boca Raton, FL: CRC.

Liggett, W, Peterson, R, and Salit, M (2008) Technical vis-à-vis biological variation in gene expression measurements. *Critical Assessment of Microarray Data Analysis*, Vienna. http://bioinf.boku.ac.at/camda/Biological%20Variability-Liggett.pdf.

Lin, DW, Coleman, IM, Hawley, S, *et al.* (2006) Influence of surgical manipulation on prostate gene expression: implications for molecular correlates of treatment effects and disease prognosis. *Journal of Clinical Oncology*, **24**, 3763–3770.

Littell, RC, Milliken, GA, Stroup, WW, and Wolfinger, RD (1996) *SAS System for Mixed Models*. Cary, NC: SAS Institute.

Liu, JS (2001) *Monte Carlo Strategies in Scientific Computing*. New York: Springer-Verlag.

Liu, X (2007) New statistical tools for microarray data and comparison with existing tools. PhD dissertation, University of North Carolina, Chapel Hill.

Liu, X, Noll, DM, Lieb, JD, and Clarke, ND (2005) DIP-chip: Rapid and accurate determination of DNA-binding specificity. *Genome Research* **15**, 412–427.

Lloyd, CM, Halstead, MD, and Nielsen, PF (2004) CellML: its future, present and past. *Progress in Biophysics and Molecular Biology*, **85**, 433–450.

Lloyd, CM, Lawson, JR, Hunter, PJ, *et al.* (2008) The CellML Model Repository. *Bioinformatics*, **24**(18), 2122–2123.

Lobenhofer, EK, Auman, JT, Blackshear, PE, *et al.* (2008) Gene expression response in target organ and whole blood varies as a function of target organ injury phenotype. *Genome Biology*, **9**, R100.

Lockhart, DJ, Dong, H, Byrne, MC, *et al.* (1996) Expression monitoring by hybridization to high-density oligonucleotide arrays. *Nature Biotechnology*, **14**, 1675–1680.

Lönnstedt, I, and Speed, TP (2002) Replicated microarray data. *Statistical Sinica*, **12**, 31–46.

Lönnstedt, I, Rimini, R and Nilsson, P (2005) Empirical Bayes microarray ANOVA and grouping cell lines by equal expression levels. *Statistical Applications in Genetics and Molecular Biology*, **4**.

Lopes, FM, Martins, DC, Jr., and Cesar, RM, Jr. (2008) Feature selection environment for genomic applications. *BMC Bioinformatics*, **9**, 451.

Lorenz, DR, Cantor, CR and Collins, JJ (2009) A network biology approach to aging in yeast. *Proceedings of the National Academy of Sciences of the USA*, **106**, 1145–1150.

Luciano, JS (2005) PAX of mind for pathway researchers. *Drug Discovery Today*, **10**, 937–942.

Lusa, L, McShane, LM, Reid, JF, *et al.* (2007) Challenges in projecting clustering results across gene expression-profiling datasets. *Journal of the National Cancer Institute*, **99**, 1715–1723.

Lynch, JL, deSilva, CJS, Peeva, VK and Swanson, NR (2006) Comparison of commercial probe labeling kits for microarray: Towards quality assurance and consistency of reactions. *Analytical Biochemistry*, **355**, 224–231.

Lyng, H, Badiee, A, Svendsrand, DH, *et al.* (2004) Profound influence of microarray scanner characteristics on gene expression ratios: analysis and procedure for correction. *BMC Genomics*, **5**, 10.

Ma, C, Lyons-Weiler, M, Liang, W, *et al.* (2006) In vitro transcription amplification and labeling methods contribute to the variability of gene expression profiling with DNA microarrays. *Journal of Molecular Diagnostics*, **8**, 183–192.

MAQC Consortium (2006) The MicroArray Quality Control (MAQC) project shows inter- and intraplatform reproducibility of gene expression measurements. *Nature Biotechnology*, **24**, 1151–1161.

Marboe, CC, Billingham, M, Eisen, H, *et al.* (2005) Nodular endocardial infiltrates (Quilty lesions) cause significant variability in diagnosis of ISHLT Grade 2 and 3A rejection in cardiac allograft recipients. *Journal of Heart and Lung Transplantation*, **24**(Suppl.), S219–S226.

Marron, JS, and Todd, MJ (2002) Distance weighted discrimination. Unpublished manuscript. http://www.optimization-online.org/DB_HTML/2002/07/513.html.

Marron, JS, Todd, MJ, and Ahn, J (2007) Distance weighted discrimination. *Journal of the American Statistical Association*, **102**, 1267–1271.

Marshall, E (2004) Getting the noise out of gene arrays. *Science*, **306**, 630–631.

Mathivanan, S, Ahmed, M, Ahn, NG, *et al.* (2008) Human Proteinpedia enables sharing of human protein data. *Nature Biotechnology*, **26**, 164–167.

McCarthy, MI, Abecasis, GR, Cardon, LR, *et al.* (2008) Genome-wide association studies for complex traits: consensus, uncertainty and challenges. *Nature Reviews Genetics*, **9**, 356–369.

McGee, M, and Chen, Z (2006) Parameter estimation for the exponential-normal convolution model for background correction of Affymetrix GeneChip data. *Statistical Applications in Genetics and Molecular Biology*, **5**.

Mecham, BH, Klus, GT, Strovel, J, *et al.* (2004) Sequence-matched probes produce increased cross-platform consistency and more reproducible biological results in microarray-based gene expression measurements. http://www.pubmedcentral.nih.gov/articlerender.fcgi?artid = 419626.

Mehta, T, Tanik, M, and Allison, DB (2004) Towards sound epistemological foundations of statistical methods for high-dimensional biology. *Nature Genetics*, **36**, 943–947.

Michiels, S, Koscielny, S, and Hill, C (2005) Prediction of cancer outcome with microarrays: a multiple random validation strategy. *Lancet*, **365**, 488–492.

Mons, B, Ashburner, M, Chichester, C, *et al.* (2008) Calling on a million minds for community annotation in WikiProteins. *Genome Biology*, **9**, R89.

Montgomery, DC (2009) *Design and Analysis of Experiments*, 7th edn. Hoboken, NJ: John Wiley & Sons, Inc.

Moreau, Y, Aerts, S, De Moor, B, *et al.* (2003) Comparison and meta-analysis of microarray data: from the bench to the computer desk. *Trends in Genetics*, **19**, 570–577.

Mullighan, CG, Goorha, S, Radtke, I, *et al.* (2007) Genome-wide analysis of genetic alterations in acute lymphoblastic leukaemia. *Nature*, **446**, 758–764.

Nakai, Y, Hashida, H, Kadota, K, *et al.* (2008) Up-regulation of genes related to the ubiquitin-proteasome system in the brown adipose tissue of 24-h-fasted rats. *Bioscience, Biotechnology and Biochemistry*, **72**, 139–148.

Nature Cell Biology (2008) Standardizing data (editorial), *Nature Cell Biology*, **10**, 1123–1124.

Newton, MA, Kendziorski, CM, Richmond, CS, *et al.* (2001) On differential variability of expression ratios: improving statistical inference about gene expression changes from microarray data. *Journal of Computational Biology*, **8**, 37–52.

Nielsen, TO, West, RB, Linn, SC, *et al.* (2002) Molecular characterisation of soft tissue tumours: a gene expression study. *Lancet*, **359**, 1301–1307.

Nguyen, DV, Arpat, AB, Wang, N, *et al.* (2002) DNA microarray experiments: biological and technological aspects. *Biometrics*, **58**, 701–717.

Novak, JP, Sladek, R, and Hudson, TJ (2002) Characterization of variability in large-scale gene expression data: implications for study design. *Genomics*, **79**, 104–113.

NTP (1996) Annual Plan for Fiscal Year 1996. Washington, DC: National Toxicology Program.

Paik, S, Shak, S, Tang, G, *et al.* (2004) A multigene assay to predict recurrence of tamoxifen-treated, node-negative breast cancer. *New England Journal of Medicine*, **351**, 2817–2826.

Palka-Santini, M, Cleven, BE, Eichinger, L, *et al.* (2009) Large scale multiplex PCR improves pathogen detection by DNA microarrays. *BMC Microbiology*, **9**, 1.

Pan, W (2005) Incorporating biological information as a prior in an empirical Bayes approach to analyzing microarray data. *Statistical Applications in Genetics and Molecular Biology*, **4**.

Parmigiani G, Garrett-Mayer ES, Anbazhagan R, *et al.* (2004) A cross-study comparison of gene expression studies for the molecular classification of lung cancer. *Clinical Cancer Research*, **10**, 2922–2927.

Pearson, TA, and Manolio, TA (2008) How to interpret a genome-wide association study. *Journal of the American Medical Association*, **299**, 1335–1344.

Pease, AC, Solas, D, Sullivan, EJ, *et al.* (1994) Light-generated oligonucleotide arrays for rapid DNA-sequence analysis. *Proceedings of the National Academy of Sciences of the USA*, **91**, 5022–5026.

Pe'er, I de, Bakker, PI, Maller, J, *et al.* (2006) Evaluating and improving power in whole-genome association studies using fixed marker sets. *Nature Genetics*, **38**, 663–667.

Petersen, D, Chandramouli, GV, Geoghegan, J, *et al.* (2005) Three microarray platforms: an analysis of their concordance in profiling gene expression. *BMC Genomics*, **6**, 63–76.

Pham, MX, Deng, MC, Kfoury, AG, *et al.* (2007) Molecular testing for long-term rejection surveillance in heart transplant recipients: design of the Invasive Monitoring Attenuation Through Gene Expression (IMAGE) trial. *Journal of Heart and Lung Transplantation*, **26**, 808–814.

Pico, AR, Kelder, T, van Iersel, MP, *et al.* (2008) WikiPathways: pathway editing for the people. *PLoS Biology*, **6**, e184.

Pinheiro, JC, and Bates, DM (2000) *Mixed-Effects Models in S and S-PLUS*. New York, Springer Verlag.

Pollack, JR, Perou, CM, Sorlie, T, *et al.* (1999) Genome-wide analysis of DNA copy number variation in breast cancer using DNA microarrays. *Nature Genetics*, **23**, 41–46.

Price, AL, Patterson, NJ, Plenge, RM, *et al.* (2006) Principal component analysis corrects for stratification in genome-wide association studies. *Nature Genetics*, **38**, 904–909.

Purvine, S, Picone, AF, and Kolker, E (2004) Standard mixtures for proteome studies. *Omics*, **8**, 79–92.

Qin, LX, and Kerr, KF (2004) Empirical evaluation of data transformations and ranking statistics for microarray analysis. *Nucleic Acids Research*, **32**, 5471–5479.

Quackenbush, J (2002) Microarray data normalization and transformation. *Nature Genetics*, **32**, 496–501.

Quackenbush, J (2004) Data standards for 'omic' science. *Nature Biotechnology*, **22**, 613–614.

Ransohoff, DF (2005) Bias as a threat to the validity of cancer molecular-marker research. *Nature Reviews Cancer*, **5**, 142–149.

Rao, Y, Lee, Y, Jarjoura, D, *et al.* (2008) A comparison of normalization techniques for MicroRNA microarray data. *Statistical Applications in Genetics and Molecular Biology*, **7**.

Rayner, TF, Rocca-Serra, P, Spellman, PT, *et al.* (2006) A simple spreadsheet-based, MIAME-supportive format for microarray data: MAGE-TAB. *BMC Bioinformatics*, **7**, 489.

Reich, DE, and Lander, ES (2001) On the allelic spectrum of human disease. *Trends in Genetics*, **17**, 502–510.

Rhodes, DR, Yu, J, Shanker, K, *et al.* (2004) Large-scale meta-analysis of cancer microarray data identifies common transcriptional profiles of neoplastic transformation and progression. *Proceedings of the National Academy of Sciences of the USA*, **101**, 9309–14.

Ritchie, ME, Silver, J, Oshlack, A, *et al.* (2007) A comparison of background correction methods for two-colour microarrays. *Bioinformatics*, **23**, 2700–2707.

Rodwell, GE, Sonu, R, Zahn, JM, *et al.* (2004) A transcriptional profile of aging in the human kidney. *PLoS Biology*, **2**, e427.

Roepman, P, Wessels, LF, Kettelarij, N, *et al.* (2005) An expression profile for diagnosis of lymph node metastases from primary head and neck squamous cell carcinomas. *Nature Genetics*, **37**, 182–186.

Ross, DT, and Perou, CM (2001) A comparison of gene expression signatures from breast tumors and breast tissue derived cell lines. *Disease Markers*, **17**, 99–109.

Ross, DT, Scherf, U, Eisen MB, *et al.* (2000) Systematic variation in gene expression patterns in human cancer cell lines. *Nature Genetics*, **24**, 227–235.

Rothman, KJ, Greenland, S, and Walker, AM (1980) Concepts of interaction. *American Journal of Epidemiology*, **112**, 467–470.

Rothman, KJ, Greenland, S, and Lash, TL (2008) *Modern Epidemiology*, 3rd edn, thoroughly rev. and updated. Philadelphia: Wolters Kluwer Health/Lippincott Williams & Wilkins.

Rudic, RD, McNamara, P, Reilly, D, *et al.* (2005) Bioinformatic analysis of circadian gene oscillation in mouse aorta. *Circulation*, **112**, 2716–2724.

Rydén, P, Andersson, H, Landfors, M, *et al.* (2006) Evaluation of microarray data normalization procedures using spike-in experiments. *BMC Bioinformatics*, **7**, 300.

Salit, M (2006) Standards in gene expression microarray experiments. *Methods in Enzymology*, **411**, 63–78.

Samani, NJ, Erdmann, J, Hall, AS, *et al.* (2007) Genomewide association analysis of coronary artery diseas. *New England Journal of Medicine*, **357**, 443–453.

Sansone, SA, and Field, D (2006) A special issue on data standards. *Omics*, **10**, 84–93.

Sansone, SA, Morrison, N, Rocca-Serra, P, *et al.* (2004) Standardization initiatives in the (eco)toxicogenomics domain: a review. *Comparative and Functional Genomics*, **5**, 633–641.

Sansone, SA, Rocca-Serra, P, Tong, W, *et al.* (2006) A strategy capitalizing on synergies: the Reporting Structure for Biological Investigation (RSBI) working group. *Omics*, **10**, 164–171.

Sansone, SA, Fan, T, Goodacre, R, *et al.* (2007) The metabolomics standards initiative. *Nature Biotechnology*, **25**, 846–848.

Sansone, SA, Rocca-Serra, P, Brandizi, M, *et al.* (2008) The first RSBI (ISA-TAB) workshop: 'can a simple format work for complex studies?'. *Omics*, **12**, 143–149.

Satterfield, MB, Lippa, K, Lu, ZQ, *et al.* (2008) Microarray scanner performance over a five-week period as measured with Cy5 and Cy3 serial dilution slides. *Journal of Research of the National Institute of Standards and Technology*, **113**, 157–174.

Saxena, R, Voight, BF, Lyssenko, V, *et al.* (2007) Genome-wide association analysis identifies loci for type 2 diabetes and triglyceride level. *Science*, **316**, 1331–1336.

Schadt, EE, Li, C, Ellis, B, *et al.* (2001) Feature extraction and normalization algorithms for high-density oligonucleotide gene expression array data. *Journal of Cellular Biochemistry*, **84**(S37), 120–125.

Scharpf, RB, Iacobuzio-Donahue, CA, Sneddon JB, *et al.* (2007) When should one subtract background fluorescence in 2-color microarrays? *Biostatistics*, **8**, 695–707.

Schaupp, CJ, Jiang, G, Myers, TG, *et al.* (2005) Active mixing during hybridization improves the accuracy and reproducibility of microarray results. *BioTechniques*, **38**, 117–119

Schena, M, Shalon, D, Davis, RW, *et al.* (1995) Quantitative monitoring of gene expression patterns with a complementary DNA microarray. *Science*, **270**, 467–470.

Schroeder, A, Mueller, O, Stocker, S, *et al.* (2006) The RIN: an RNA integrity number for assining integrity values to RNA measurements. *BMC Molecular Biology*, **7**, 3.

Scott, LJ, Mohlke, KL, Bonnycastle, LL, *et al.* (2007) A genome-wide association study of type 2 diabetes in Finns detects multiple susceptibility variants. *Science*, **316**, 1341–1345.

Shabalin, AA, Tjelmeland, H, Fan, C, *et al.* (2008) Merging two gene-expression studies via cross-platform normalization. *Bioinformatics*, **24**, p. 1154–60.

Shi, L, Tong, W, Fang, H, *et al.* (2005a) Cross-platform comparability of microarray technology: intra-platform consistency and appropriate data analysis procedures are essential. *BMC Bioinformatics*, **6**, S12.

Shi, L, Tong, W, Su, Z, *et al.* (2005b) Microarray scanner calibration curves: characteristics and implications. *BMC Bioinformatics*, **6**, S11.

Shi, L, Reid, LH, Jones, WD, *et al.* (2006) The MicroArray Quality Control (MAQC) project shows inter- and intraplatform reproducibility of gene expression measurements. *Nature Biotechnology*, **24**, 1151–1161.

Shi, L, Perkins, RG, Fang, H, *et al.* (2008) Reproducible and reliable microarray results through quality control: good laboratory proficiency and appropriate data analysis practices are essential. *Current Opinion in Biotechnology*, **19**, 10–18.

Shippy, R, Sendera, TJ, Lockner, R, *et al.* (2004) Performance evaluation of commercial short-oligonucleotide microarrays and the impact of noise in making cross-platform correlations. *BMC Genomics*, **5**, 61.

Sica, GT (2006) Bias in research studies. *Radiology*, **238**, 780–789.

Simon, R (2005) Roadmap for developing and validating therapeutically relevant genomic classifiers. *Journal of Clinical Oncology*, **23**, 7332–7341.

Simpson, EH (1951) The interpretation of interaction in contingency tables. *Journal of the Royal Statistical Society, Series B*, **13**, 238–241.

Sims, AH, Smethurst, GJ, Hey, Y., *et al.* (2008) The removal of multiplicative, systematic bias allows integration of breast cancer gene expression datasets – improving meta-analysis and prediction of prognosis. *BMC Medical Genomics*, **1**, 42.

Singh-Gasson, S, Green, RD, Yue, Y, *et al.* (1999) Maskless fabrication of light-directed oligonucleotide microarrays using a digital micromirror array. *Nature Biotechnology*, **17**, 974–978.

Sladek, R, Rocheleau, G, Rung, J, *et al.* (2007) A genome-wide association study identifies novel risk loci for type 2 diabetes. *Nature*, **445**, 881–885.

Smith, B, Ashburner, M, Rosse, C, *et al.* (2007) The OBO Foundry: coordinated evolution of ontologies to support biomedical data integration. *Nature Biotechnology*, **25**, 1251–1255.

Smyth, GK (2004) Linear models and empirical Bayes methods for assessing differential expression in microarray experiments. *Statistical Applications in Genetics and Molecular Biology*, **3**.

Soltis, DE, Ma, H, Frohlich, MW, *et al.* (2007) The floral genome: an evolutionary history of gene duplication and shifting patterns of gene expression. *Trends in Plant Science*, **12**, 358–367.

Souza, T, Kush, R, and Evans, JP (2007) Global clinical data interchange standards are here!' *Drug Discovery Today*, **12**, 174–181.

Spellman, PT, Miller, M, Stewart, J, *et al.* (2002) Design and implementation of microarray gene expression markup language (MAGE-ML). *Genome Biology*, **3**, research0046.

Steinthorsdottir, V, Thorleifsson, G, Reynisdottir, I, *et al.* (2007) A variant in CDKAL1 influences insulin response and risk of type 2 diabetes. *Nature Genetics*, **39**, 770–775.

Stafford, P (ed.) (2008) *Methods in Microarray Normalization*. Boca Raton, FL: CRC.

Stein, LD (2008) Towards a cyberinfrastructure for the biological sciences: progress, visions and challenges. *Nature Review Genetics*, **9**, 678–688.

Strauss, E (2006) Arrays of hope. *Cell*, **127**, 657–659.

Stolovitzky, G (2003) Gene selection in microarray data: the elephant, the blind men and our algorithms. *Currenr Opinion in Structural Biology*, **13**, 370–376.

Storey, JD, and Tibshirani, R (2003) Statistical significance for genomewide studies. *Proceedings of the National Academy of Sciences of the USA*, **100**, 9440–9445.

Tan, PK, Downey, TJ, Spitznagel, EL Jr, *et al.* (2003) Evaluation of gene expression measurements from commercial microarray platforms. *Nucleic Acids Research*, **31**, 5676–5684.

Taylor, CF, Hermjakob, H, Julian, RK, Jr, *et al.* (2006) The work of the Human Proteome Organisation's Proteomics Standards Initiative (HUPO PSI). *Omics*, **10**, 145–151.

Taylor, CF, Paton, NW, Lilley, KS, *et al.* (2007) The minimum information about a proteomics experiment (MIAPE). *Nature Biotechnology*, **25**, 887–893.

Taylor, CF, Field, D, Sansone, SA, *et al.* (2008) Promoting coherent minimum reporting guidelines for biological and biomedical investigations: the MIBBI project. *Nature Biotechnology*, **26**, 889–896.

Tchetcherina, N (2007) ANCOVA based normalization method for two-channel microarrays. MS thesis, Pennsylvania State University.

Thomas, DC, and Witte, JS (2002) Point: Population stratification: a problem for case-control studies of candidate-gene associations? *Cancer Epidemiology, Biomarkers and Prevention*, **11**, 505–512.

Thomas, RS, O'Connell, TM, Pluta, L, *et al.* (2007a) A comparison of transcriptomic and metabonomic technologies for identifying biomarkers predictive of two-year rodent cancer bioassays. *Toxicological Sciences*, **96**, 40–46.

Thomas, RS, Pluta, L, Yang, L, *et al.* (2007b) Application of genomic biomarkers to predict increased lung tumor incidence in 2-year rodent cancer bioassays. *Toxicological Sciences*, **97**, 55–64.

Thompson, KL, Pine, PS, Rosenzweig, BA, *et al.* (2007) Characterization of the effect of sample quality on high density oligonucleotide microarray data using progressively degraded rat liver RNA. *BMC Biotechnology* **7**, 57.

Tibshirani, R, Hastie, T, Narasimhan, B, *et al.* (2002) Diagnosis of multiple cancer types by shrunken centroids of gene expression. *Proceedings of the National Academy of Sciences of the USA*, **99**, 6567–6572.

Troyanskaya, O, Cantor, M, Sherlock, G, *et al.* (2001) Missing value estimation methods for DNA microarrays. *Bioinformatics*, **17**, 520–525.

Trygg, J and Wold, S (2002) Orthogonal projections to latent structures (O-PLS). *Journal of Chemometrics*, **16**, 119–128.

Tseng, GC, Oh, MK, Rohlin, L, *et al.* (2001) Issues in cDNA microarray analysis: quality filtering, channel normalization, models of variations and assessment of gene effects. *Nucl.Acids Research*, **29**, 2549–57.

Tu, Y, Stolovitzky, G, and Klein, U, (2002) Quantitative noise analysis for gene expression microarray experiments. *Proceedings of the National Academy of Sciences of the USA*, **99**, 14031–14036.

Tukey, J, and Tukey, P (1990) Strips displaying empirical distributions: textured dot strips. *Bellcore Technical Memorandum*.

Tusher, VG, Tibshirani, R and Chu, G (2001) Significance analysis of microarrays applied to the ionizing radiation response. *Proceedings of the National Academy of Sciences of the USA*, **98**, 5116–5121.

United States Environmental Protection Agency (2007) *Guidance for Preparing Standard Operating Procedures*, EPA QA/G-6, Washington, DC: EPA. http://www.epa.gov//quality1/qs-docs/g6-final.pdf.

Upton, GJ, Langdon, WB, and Harrison, AP (2008) G-spots cause incorrect expression measurement in Affymetrix microarrays. *BMC Genomics*, **9**, 613.

Van Bakel, H, and Holstege, FCP (2004) In control: systematic assessment of microarray performance. *EMBO Reports*, **5**, 964–969.

Van Gelder, RN, von Zastrow, ME, Yool, A (1990) Amplified RNA synthesized from limited quantities of heterogeneous cDNA. *Proceedings of the National Academy of Sciences of the USA*, **87**, 1663–1667.

Vastrik, I, D'Eustachio, P, Schmidt, E, *et al.* (2007) Reactome: a knowledge base of biologic pathways and processes. *Genome Biology*, **8**, R39.

Velculescu, VE, Zhang, L, Vogelstein, B, *et al.* (1995) Serial analysis of gene expression. *Science*, **270**, 484–487.

Venter, JC, Adams, MD, Myers, EW, *et al.* (2001) The sequence of the human genome. *Science*, **291**, 1304–1351.

Wacholder, S, Rothman, N, and Caporaso, N (2002) Counterpoint: Bias from population stratification is not a major threat to the validity of conclusions from epidemiological studies of common polymorphisms and caner. *Cancer Epidemiology, Biomarkers and Prevention*, **11**, 513–520.

Walker, WL, Liao, IH, Gilbert, DL, *et al.* (2008) Empirical Bayes accomodation of batch-effects in microarray data using identical replicate reference samples: application to RNA expression profiling of blood from Duchenne muscular dystrophy patients. *BMC Genomics*, **9**, 494.

Wang, Y, Barbacioru, C, Hyland, F, *et al.* (2006) Large scale real-time PCR validation on gene expression measurements from two commercial long-oligonucleotide microarrays. *BMC Genomics*, **7**, 59.

Waring, JF, Ulrich, RG, Flint, N, *et al.* (2004) Interlaboratory evaluation of rat hepatic gene expression changes induced by methapyrilene genomics and risk assessment. *Environmental Health Perspectives*, **112**, 439–448.

Wellcome Trust Case Control Consortium (2007) Genome-wide association study of 14,000 cases of seven common diseases and 3,000 shared controls. *Nature*, **447**, 661–678.

Whetzel, PL, Brinkman, RR, Causton, HC, *et al.* (2006a) Development of FuGO: an ontology for functional genomics investigations. *Omics*, **10**, 199–204.

Whetzel, PL, Parkinson, H, Causton, HC, *et al.* (2006b) The MGED Ontology: a resource for semantics-based description of microarray experiments. *Bioinformatics*, **22**, 866–873.

Whiting, P, Rutjes, AWS, Reitsma, JB, *et al.* (2004) Sources of variation and bias in studies of diagnostic accuracy. *Annals of Internal Medicine*, **140**, 189–202.

Whitney, AR, Diehn, M, Popper, SJ, *et al.* (2003) Individuality and variation in gene expression patterns in human blood. *Proceedings of the National Academy of Sciences of the USA*, **100**, 1896–1901.

Wilson, DL, Buckley, MJ, Helliwell, CA, *et al.* (2003) New normalization methods for cDNA microarray data. *Bioinformatics*, **19**, 1325–1332.

Wit, E, and McClure, J (2004) *Statistics for Microarrays*. Chichester: John Wiley & Sons, Ltd.

Wolfinger, RD, Gibson, G, Wolfinger, ED, *et al.* (2001) Assessing gene significance from cDNA microarray expression data via mixed models. *Journal of Computational Biology*, **8**, 625–637.

Wooding, WM (1994) *Planning Pharmaceutical Clinical Trials*. New York: John Wiley & Sons, Inc.

Wright, GW, and Simon, RM (2003) A random variance model for detection of differential gene expression in small microarray experiments. *Bioinformatics*, **19**, 2448–2455.

Yang, YH, and Speed, TP (2002) Design issues for cDNA microarray experiments. *Nature Reviews Genetics*, **3**, 579–588.

Yang, YH, and Thorne, NP (2003) Normalization for two-color cDNA microarray data. In DR Goldstein (ed.), *Science and Statistics: A Festschrift for Terry Speed*, pp. 403–418. Beachwood, OH: Institute of Mathematical Statistics.

Yang, YH, Dudoit, S, Luu, P, *et al.* (2001) Normalization for cDNA microarray data. In ML Bittner, Y Chen, AN Dorsel and ER Dougherty (eds), *Proceedings of SPIE*, 4266, 141–152.

Yang, YH, Buckley, MJ, Dudoit, S, *et al.* (2002a) Comparison of methods for image analysis on cDNA microarray data. *Journal of Computational and Graphical Statistics*, **11**, 108–136.

Yang, YH, Buckley, MJ, and Speed, TP (2002b) Analysis of cDNA microarray images. *Briefings in Bioinformatics*, **2**, 341–349.

Yang YH, Dudoit S, Luu P, *et al.* (2002c) Normalization for cDNA microarray data: a robust composite method addressing single and multiple slide systematic variation. *Nucleic Acids Research*, **30**, e15.

Ye, QH, Qin, LX, Forgues, M, *et al.* (2003) Predicting hepatitis B virus-positive metastatic hepatocellular carcinomas using gene expression profiling and supervised machine learning. *Nature Medicine*, **9**, 416–423.

Yeoh, EJ, Ross ME, Shurtleff, SA, *et al.* (2002) Classification, subtype discovery, and prediction of outcome in pediatric acute lymphoblastic leukemia by gene expression profiling. *Cancer Cell*, **1**, 133–143.

Yin, W, Chen, T, Zhou, SX, *et al.* (2005) Background correction for cDNA microarray images using the TV+L1 model. *Bioinformatics*, **21**, 2410–2416.

Ying, L, and Sarwal, M (2009) In praise of arrays. *Pediatric Nephrology*, **24**, 1643–1659.

Zakharkin, SO, Kim, K, Mehta, T, *et al.* (2005) Sources of variation in Affymetrix microarray experiments. *BMC Bioinformatics*, **6**, 214.

Zeggini, E, Weedon, MN, Lindgren, CM, *et al.* (2007) Replication of genome-wide association signals in UK samples reveals risk loci for type 2 diabetes. *Science*, **316**, 1336–1341.

Zervakis, M, Blazadonakis, ME, Tsiliki, G, *et al.* (2009) Outcome prediction based on microarray analysis: a critical perspective on methods. *BMC Bioinformatics*, **10**, 53.

Zhang, W, Shmulevich, I, and Astola, J (2004) *Microarray Quality Control*. Hoboken NJ: Wiley-Lyss.

Zhang, X, Feng, B, Zhang, Q, *et al.* (2005) Genome-wide expression profiling and identification of gene activities during early flower development in *Arabidopsis*. *Plant Molecular Biology*, **58**, 401–19

Zheng, G, Freidlin, B, and Gastwirth, JL (2006) Robust genomic control for association studies. *American Journal of Human Genetics*, **78**, 350–356.

Zien, A, Aigner, T, Zimmer, R, *et al.* (2001) Centralization: a new method for the normalization of gene expression data. *Bioinformatics*, **17**, S323–S331.

Zimmermann, P, Schildknecht, B, Craigon, D, *et al.* (2006) MIAME/Plant adding value to plant microarray experiments. *Plant Methods*, **2**, 1–3.

Index

Batch Effects and Noise in Microarray Experiments: Sources and Solutions edited by A. Scherer
© 2009 John Wiley & Sons, Ltd